Learn to Use Microsoft Excel 2016

Michelle N. Halsey, PMP, CSM

ISBN-10: 1-64004-257-1
ISBN-13: 978-1-64004-257-5

Silver City Publications & Training, L.L.C.
P.O. Box 1914
Nampa, ID 83653

https://www.silvercitypublications.com/shop/

Contents

Chapter 1 – Opening Excel

Welcome to the new and improved Microsoft Excel 2016. This chapter will teach you how to open Microsoft Excel files and create new ones. First, we will learn how to open Microsoft Excel. You will learn how to open files from the Recent list or other files. Then you will learn how to create a blank workbook or a workbook from a template.

Opening Excel

To open Excel use the following procedure.

Step 1: From the Start page, select the Excel 2016 icon.

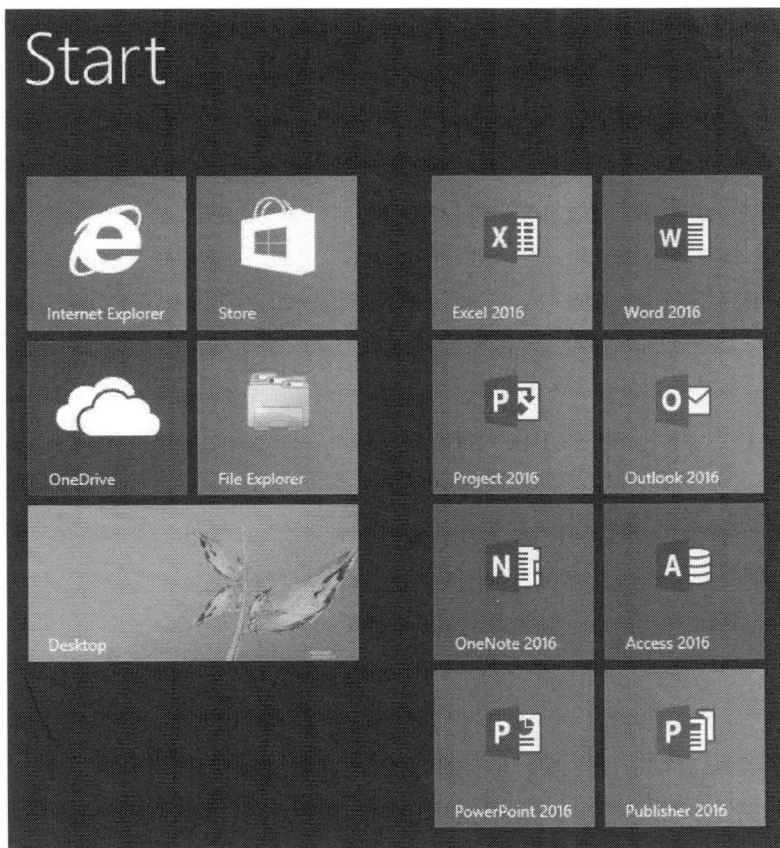

You can get to this screen by pressing the Windows key.

Using the Recent List

To open a workbook from the Recent list, use the following procedure.

Step 1: Select the workbook that you want to open from the Recent list.

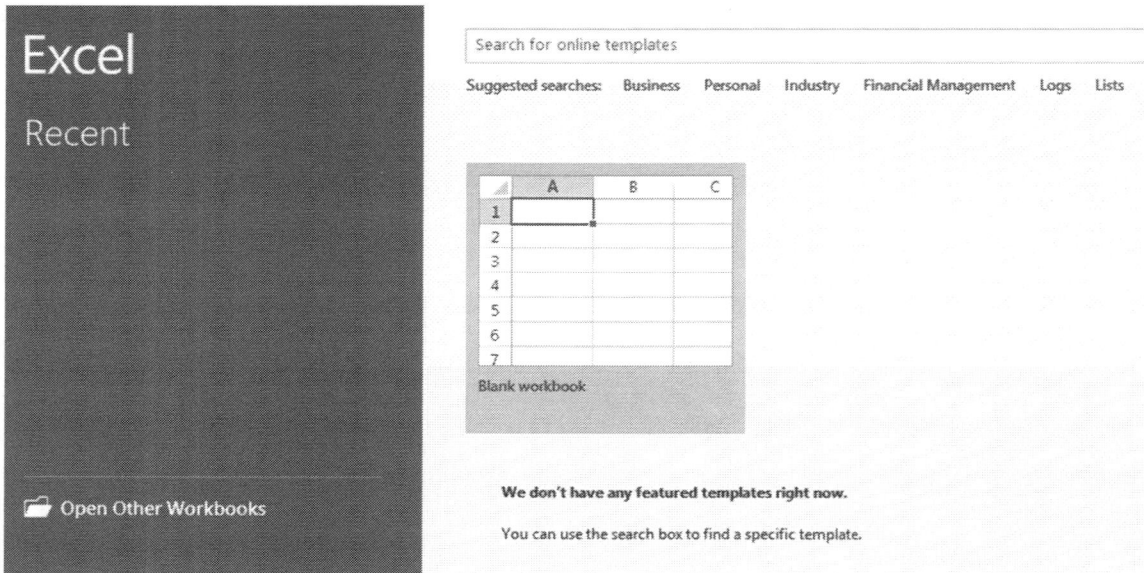

To pin an item on the Recent list, use the following procedure.

Step 1: Click the pin on the right side of the Recent list item. The item moves to the top section of the Recent list.

To unpin an item, click the pin on the right side of the Recent list again. The item returns to the previous location in the Recent list.

Opening Files

To open a workbook, use the following procedure.

Step 1: Select File and then Open from the Backstage View.

Step 2: Select one of the Places you would like to look for the workbook. The default options are Recent Workbooks, Shared with Me, Microsoft OneDrive location, This PC, Add a Place, and Browse.

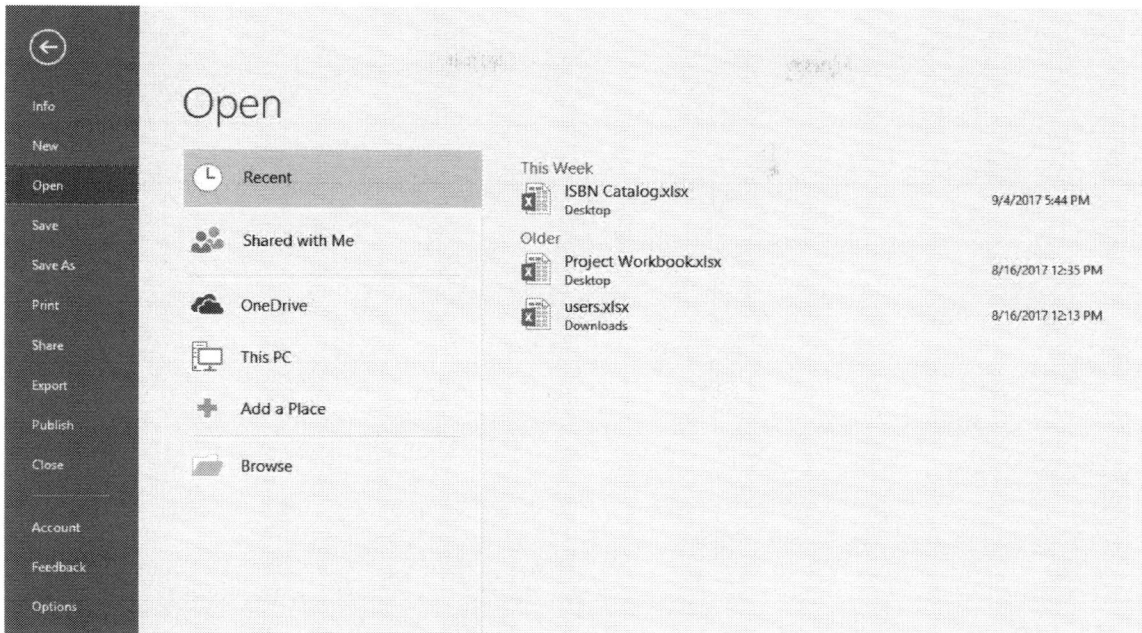

To open a document from the OneDrive or your computer, select Browse.

In the Open dialog box, navigate to the location of the file you want to open. Select it and click Open.

Creating a Blank Workbook

To create a blank workbook, use the following procedure.

Step 1: If the Backstage view is not showing, select the File tab from the Ribbon. Select New.

Step 2: From the New tab, or if you have just opened Excel 2016, select Blank Workbook.

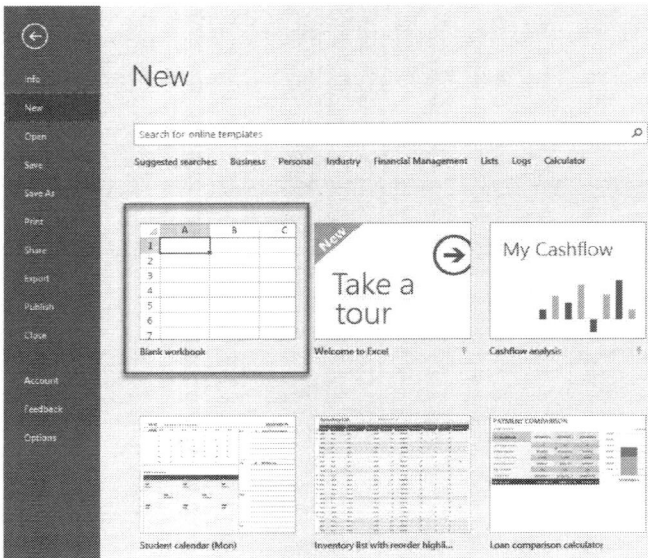

Creating a Workbook from a Template

To create a blank workbook from a template, use the following procedure.

Step 1: If the Backstage view is not showing, select the File tab from the Ribbon. Select New.

Step 2: From the New tab, or if you have just opened Excel 2016, select the template you want to use.

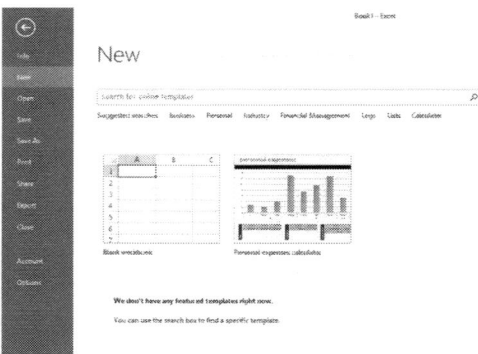

Step 3: Select Create.

You can use the left and right arrows to review the other templates in the current search.

To search for a template and filter the results, use the following procedure.

Step 1: Select one of the Suggested Search terms or enter a term in the Search box and press Enter.

Step 2: To apply a filter, select the Filter term from the list on the right side of the screen.

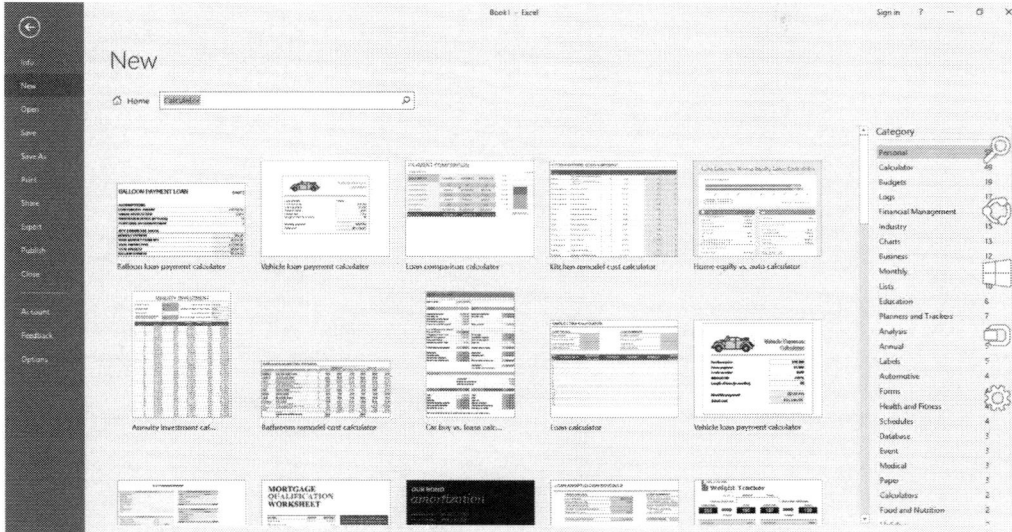

Step 3: To return to the list of templates, select Home.

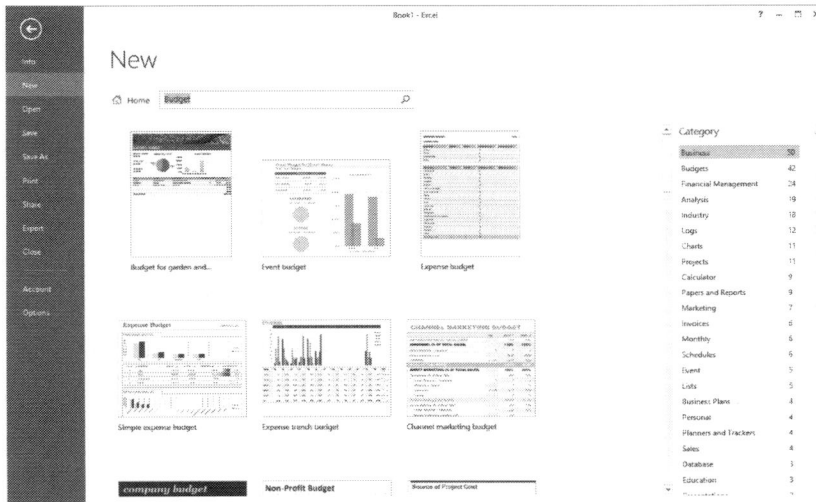

Chapter 2 – Working with the Interface

In this chapter, we will introduce you to the Excel 2016 interface, which uses the Ribbon from the previous two versions of Excel. You will get a closer look at the Ribbon and the Status bar. You will also learn how to manage your Microsoft account right from a new item above the Ribbon. This chapter introduces you to the Backstage view, where all the functions related to your files live. You will learn how to save files. Finally, we will look at closing files and closing the application.

Understanding the Interface

Explore the Excel interface, including the Ribbon, the formula bar, the worksheet area, the Quick Access toolbar, and the Status bar.

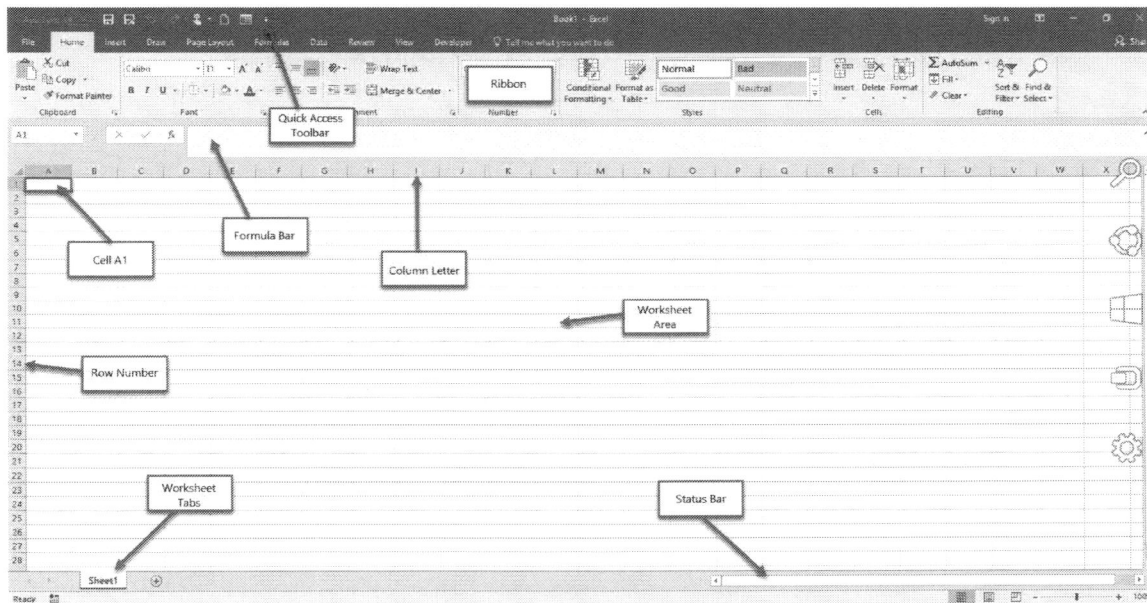

Each Tab in the Ribbon contains many tools for working with your workbook. To display a different set of commands, click the Tab name. Buttons are organized into groups according to their function.

The Quick Access toolbar appears at the top of the Excel window. It provides you with one-click shortcuts to commonly used functions, like save, undo, and redo.

We will discuss the Formula bar more in a later chapter.

The Status bar shows if any macros are currently running. It also allows you to quickly change your view or zoom of the workbook.

To zoom in or out, use the following procedure.

Step 1: Click the minus sign in the Status bar to zoom out. Click the plus sign in the Status bar to zoom in. You can also drag the slider to adjust the zoom.

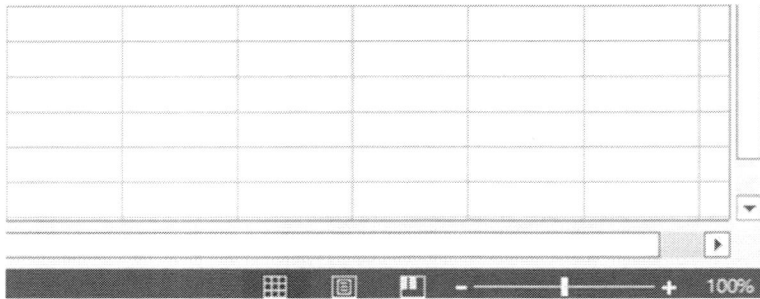

You can also click on the number percentage to open the Zoom dialog box.

About Your Account and Feedback

Explore the account options, use the following procedure.

Step 1: Click the File ribbon.

Step 2: Click the Feedback option on the menu.

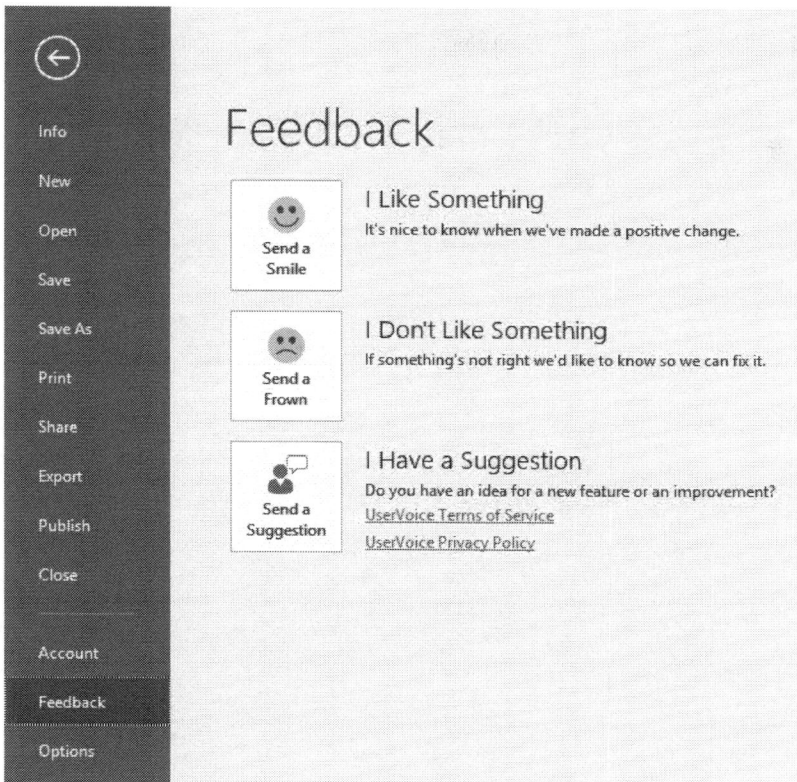

Step 3: Select the feedback option and provide the feedback.

Step 4: Enter the information requested in the Microsoft Office Feedback dialog. Select Submit.

Explore the Backstage View, use the following procedure.

Step 1: Select the File tab on the Ribbon.

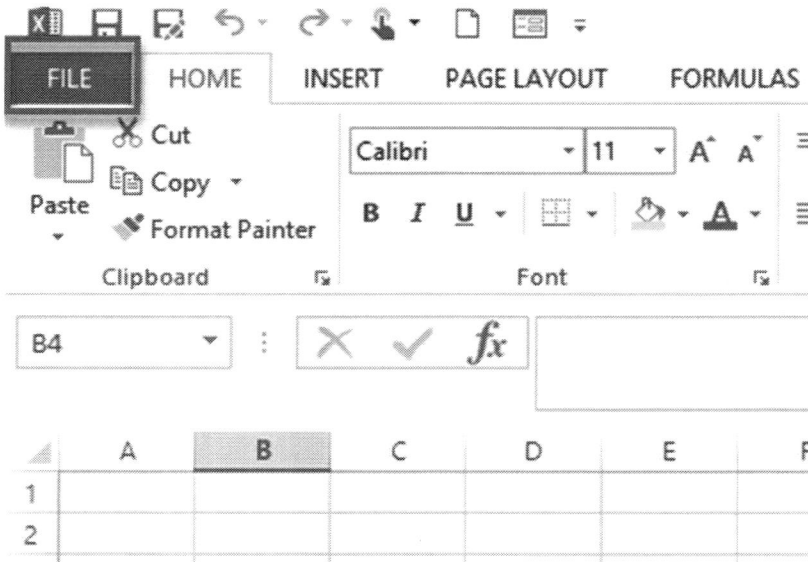

Excel displays the Backstage View, open to the Info tab by default. A sample is illustrated below.

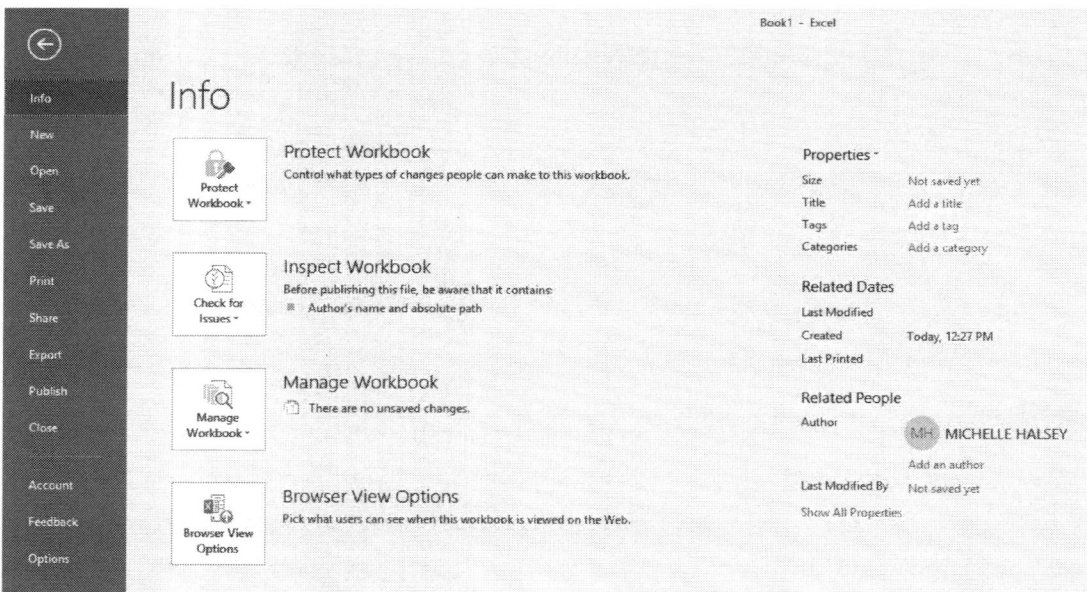

To move the active cell, point out the highlighted row and column for the active cell, as well as the name of that cell in the Name box.

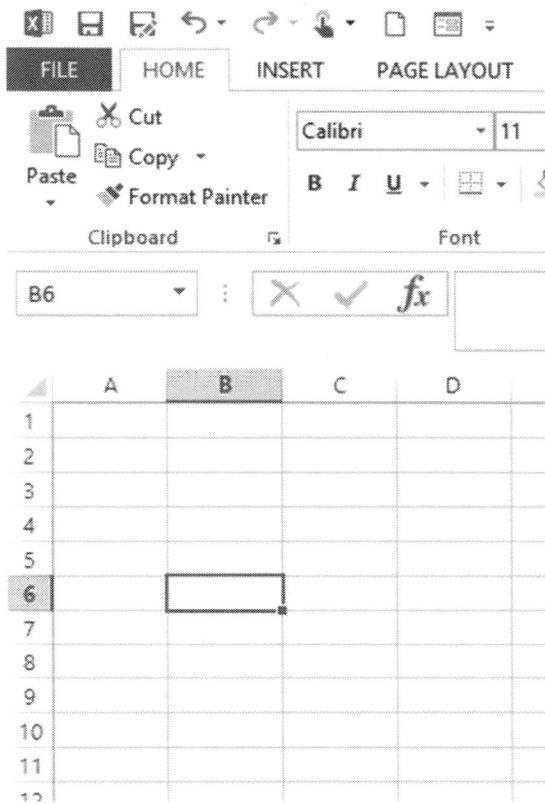

To insert a new worksheet, use the following procedure.

Step 1: Click the New Sheet plus sign.

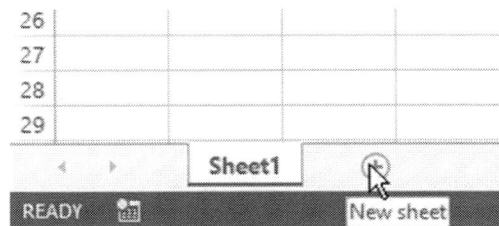

Excel opens the new worksheet to the first cell, so that you can begin entering data right away. You can rename the worksheet if needed.

To rename a worksheet, use the following procedure.

Step 1: Right click on the sheet tab and select Rename from the context menu.

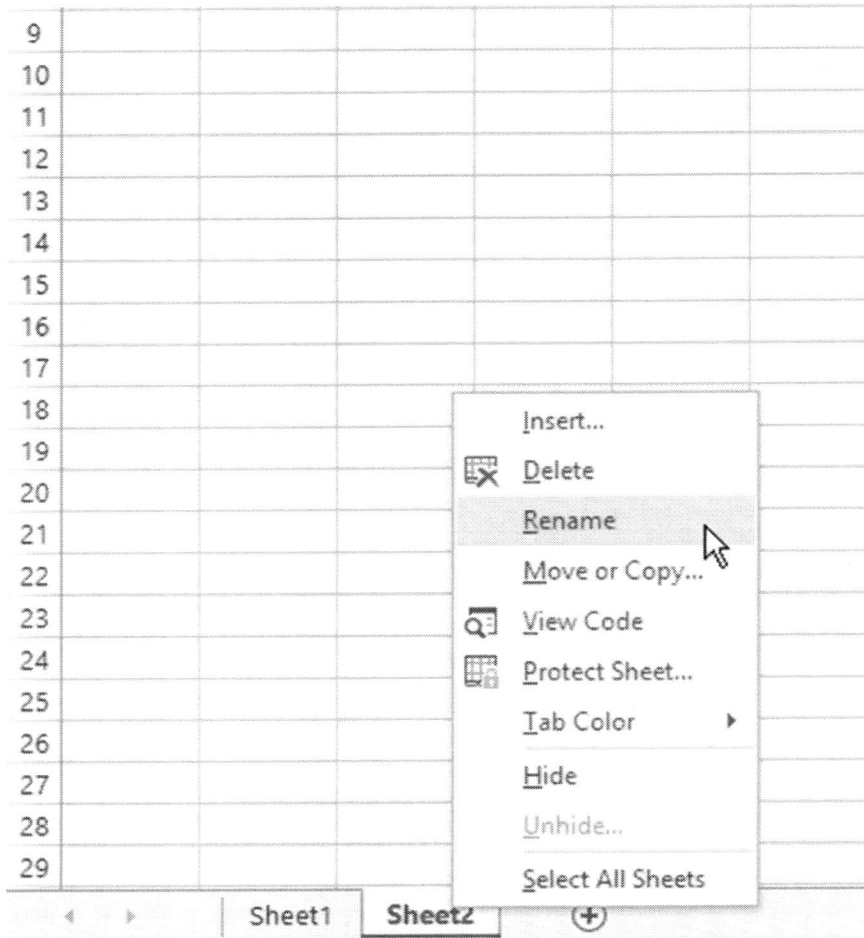

9	
10	
11	
12	
13	
14	
15	
16	
17	
18	Insert...
19	Delete
20	Rename
21	
22	Move or Copy...
23	View Code
24	Protect Sheet...
25	Tab Color ▶
26	
27	Hide
28	Unhide...
29	Select All Sheets

Sheet1 Sheet2 ⊕

Step 2: Enter the new name over the highlighted text.

To switch to a different worksheet, use the following procedure.

Step 1: Click on the worksheet tab that you want to view.

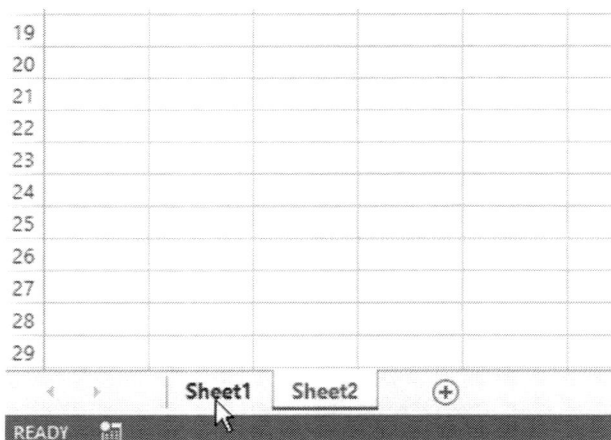

19
20
21
22
23
24
25
26
27
28
29

Sheet1 Sheet2 ⊕

READY

To move a worksheet, use the following procedure.

Step 1: Click on the worksheet tab that you want to move and drag it to the new location in the workbook.

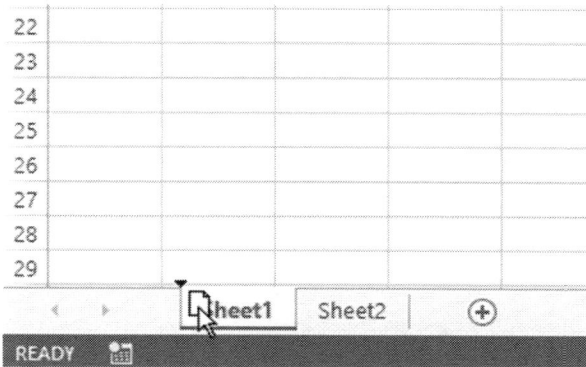

Saving Files

To save a workbook that has not been previously saved, use the following procedure.

Step 1: Select the File tab on the Ribbon.

Step 2: Select the Save command in the Backstage View.

Step 3: Select the Place where you want to save the workbook.

Step 4: If you choose your OneDrive, you can select the Documents folder. If you choose your This PC, select your Current Folder or one of your Recent Folders. Or in either place, you can choose Browse to select a new location.

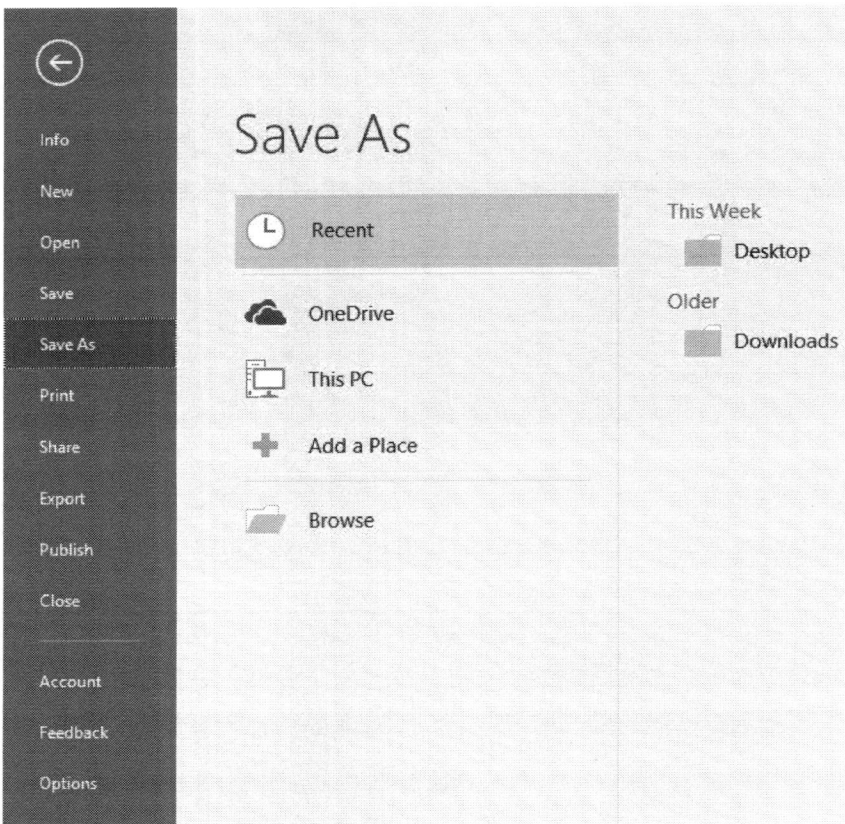

Step 5: The Save As dialog opens. Enter a File Name, and if desired, navigate to a new location to store the file. Select Save.

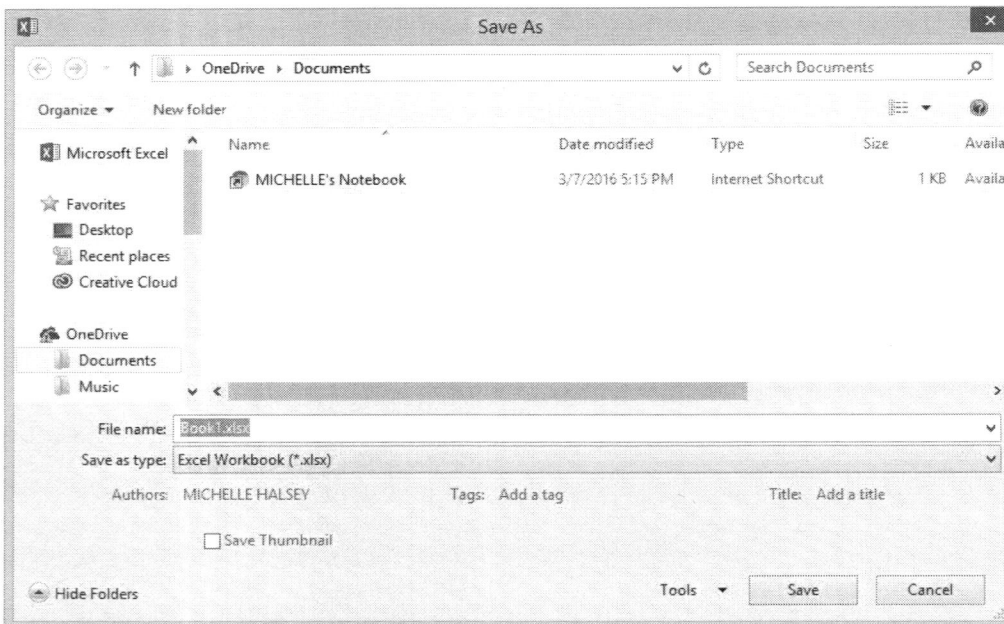

To close a file, use the following procedure.

Step 1: Select the File tab from the Ribbon.

Step 2: Select Close from The Backstage View.

If you have not saved your file, you will see the following message.

To close the application (if only one workbook is open), use the following procedure.

Step 1: Click the X at the top right corner of the window.

Chapter 3 – Your First Worksheet

In this chapter, you will start entering data into a worksheet, including using flash fill and auto fill to quickly populate the information you need to store in your worksheet. You will also learn about editing data, including checking your spelling. Since you probably do not want to move all those rows or columns when you realize that you forgot one, you will also learn how to add rows and columns.

Entering Data

Review how a long label will overlap to the next column. In the following example, "Car Payment" is too long for the column width.

	A	B	C	D	E	F	G	H	I	J
1	Household Budget									
2	2013									
3										
4		January	February	March	April	May	June	Total - first six months		
5	Mortgage	890	890	890	890	890	890			
6	Heat	250	250	175	125	80	0			
7	Power	225	225	175	175	150	150			
8										
9	Phone	65	75	65	65	75	75			
10	Car Paym	275	275	275	275	275	275			
11	Gas	240	240	360	240	240	240			
12	Insrance	180	180	180	180	180	180			
13	Food	600	600	600	600	600	600			
14										
15										
16										
17										

To widen a column, use the following procedure.

Step 1: Click on the column you want to widen. Notice the cursor changes to a cross with double arrows. The screen tips indicate how wide in pixels the column currently is.

Step 2: Drag the border to the new width

Using Auto Fill

To create a list using AutoFill, use the following procedure. This example creates new columns in the Budget worksheet to cover the second six months.

Step 1: Create a new column heading with the text "July" in cell J4.

Step 2: Select that cell to make it active. Excel displays a handle around the cell.

Step 3: Drag the handle across the columns. Excel displays a screen tip showing what AutoFill will place in those cells.

Step 4: Release the mouse button at the end of the range.

Editing Data

Explore the relationship between the active cell and the Formula Bar.

Checking Your Spelling

Explore the Spell Checker, use the following procedure.

Step 1: Select the Spelling tool on the Review tab of the Ribbon.

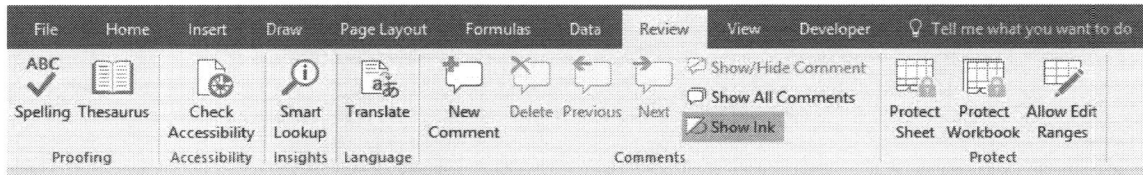

Excel opens the Spelling dialog box and begins indicating any spelling errors.

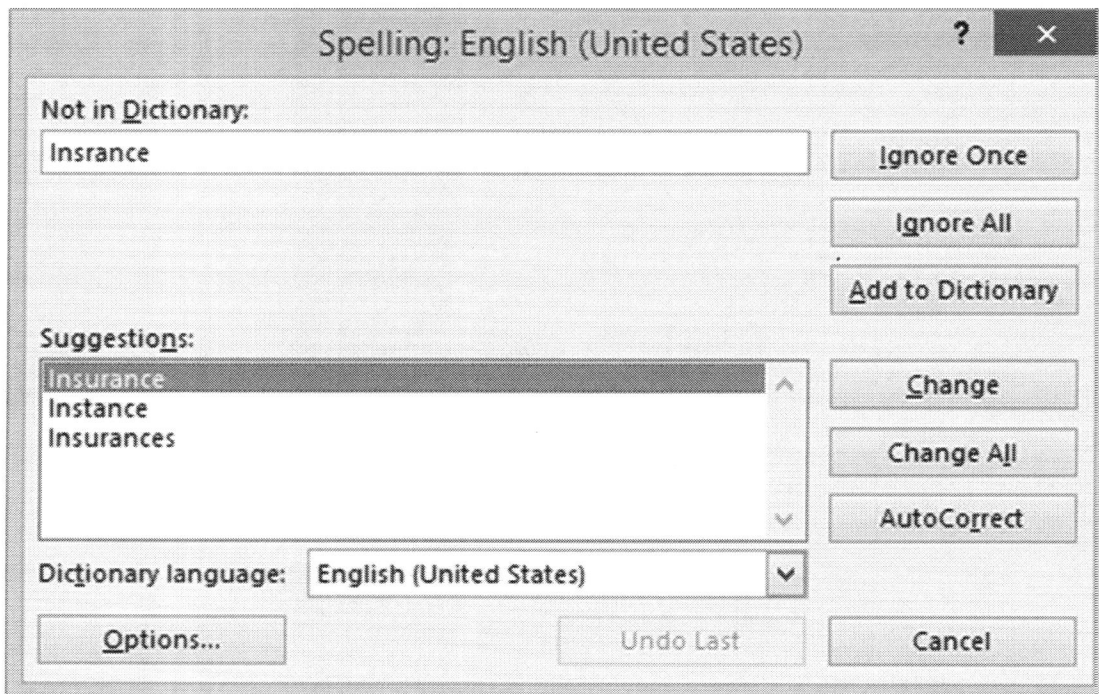

Step 2: Highlight the correct spelling and select Change. If the correct spelling is not listed, you can correct the spelling by editing the text in the Not in Dictionary field.

Adding Rows and Columns

To add a new row, use the following procedure.

Step 1: Highlight the row below where you want to insert a row. Click to the left of the row number to highlight the whole row.

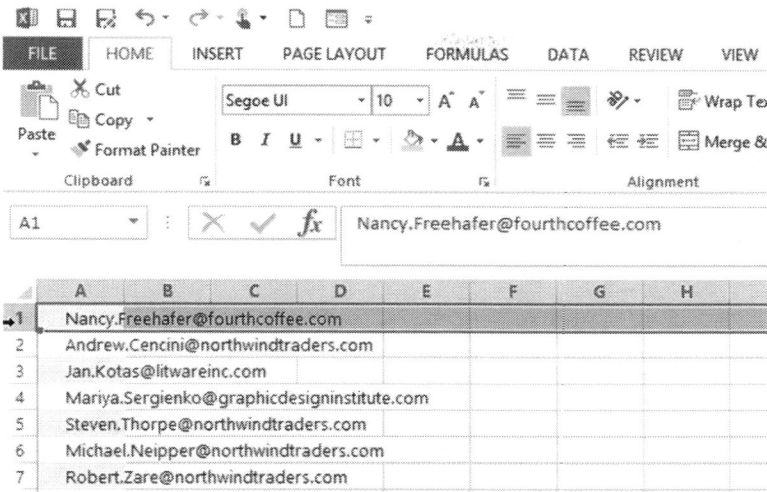

Step 2: Select Insert Sheet Rows from the Ribbon.

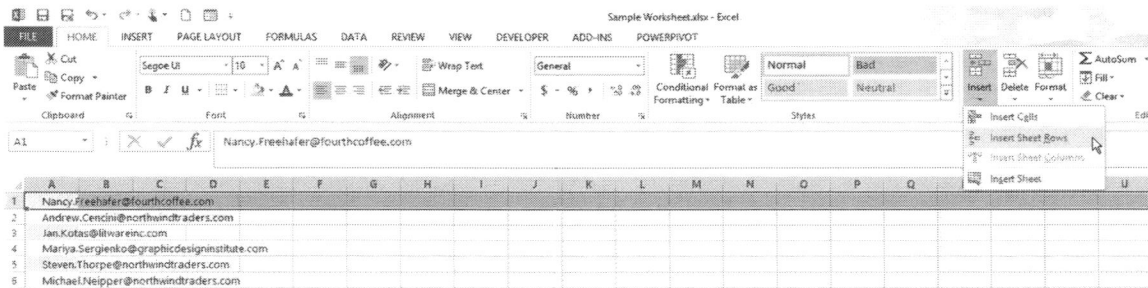

Using Flash Fill

To use flash fill, use the following procedure.

Step 1: Enter Nancy in the First Name cell of the first row.

Step 2: Enter Andrew in the First Name cell of the second row.

Step 3: When Excel displays the remaining list, press Enter to accept.

▲	A	B	C
1	Email Address	First Name	Last Name
2	Nancy.Freehafer@fourthcoffee.com	Nancy	
3	Andrew.Cencini@northwindtraders.com	Andrew	
4	Jan.Kotas@litwareinc.com	Jan	
5	Mariya.Sergienko@graphicdesigninstitute.com	Mariya	
6	Steven.Thorpe@northwindtraders.com	Steven	
7	Michael.Neipper@northwindtraders.com	Michael	
8	Robert.Zare@northwindtraders.com	Robert	
9	Laura.Giussani@adventure-works.com	Laura	
10	Anne.HL@northwindtraders.com	Anne	
11	Alexander.David@contoso.com	Alexander	
12	Kim.Shane@northwindtraders.com	Kim	
13	Manish.Chopra@northwindtraders.com	Manish	
14	Gerwald.Oberleitner@northwindtraders.com	Gerwald	
15	Amr.Zaki@northwindtraders.com	Amr	
16	Yvonne.McKay@northwindtraders.com	Yvonne	
17	Amanda.Pinto@northwindtraders.com	Amanda	
18			
19			

Step 4: Notice the icon that appears with a context menu of additional options.

▲	A	B	C
1	Email Address	First Name	Last Name
2	Nancy.Freehafer@fourthcoffee.com	Nancy	
3	Andrew.Cencini@northwindtraders.com	Andrew	
4	Jan.Kotas@litwareinc.com	Jan	
5	Mariya.Sergienko@graphicdesigninstitute.com	Mariya	
6	Steven.Thorpe@northwindtraders.com	✓ Accept suggestions	
7	Michael.Neipper@northwindtraders.com	Select all 0 blank cells	
8	Robert.Zare@northwindtraders.com	Select all 0 changed cells	
9	Laura.Giussani@adventure-works.com		
10	Anne.HL@northwindtraders.com	Anne	
11	Alexander.David@contoso.com	Alexander	
12	Kim.Shane@northwindtraders.com	Kim	
13	Manish.Chopra@northwindtraders.com	Manish	
14	Gerwald.Oberleitner@northwindtraders.com	Gerwald	
15	Amr.Zaki@northwindtraders.com	Amr	
16	Yvonne.McKay@northwindtraders.com	Yvonne	
17	Amanda.Pinto@northwindtraders.com	Amanda	
18			

Chapter 4 – Viewing Excel Data

Excel offers several options for viewing your worksheets. This chapter will provide an overview of the different views that are available. It also explains how to switch views and create a custom view. This chapter covers how to use the Zoom feature. Finally, this chapter discusses how to switch between different open files.

An Overview of Excel's Views

Explain the different view options in Excel.

- Normal is the view used for entering data.

- Page Break Preview allows you to adjust where the page breaks occur. You can drag the blue border to a new location for columns or rows to adjust the page breaks.

- Page Layout view displays what the data will look like when printed. You can use Page Layout view to add headers and footers to your worksheets.

Explore the View tab on the Ribbon.

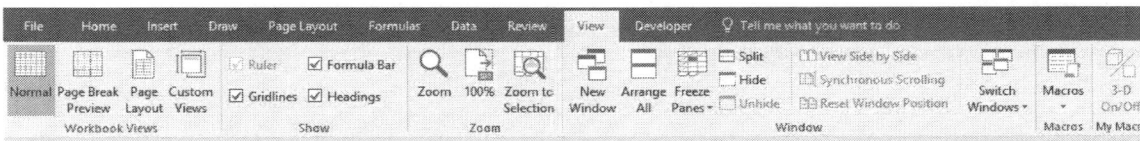

Explore the view options on the Status bar.

Switching Views

Explore the Page Break Preview, use the following procedure.

Step 1: Select the View tab from the Ribbon.

Step 2: Select the Page Break Preview tool.

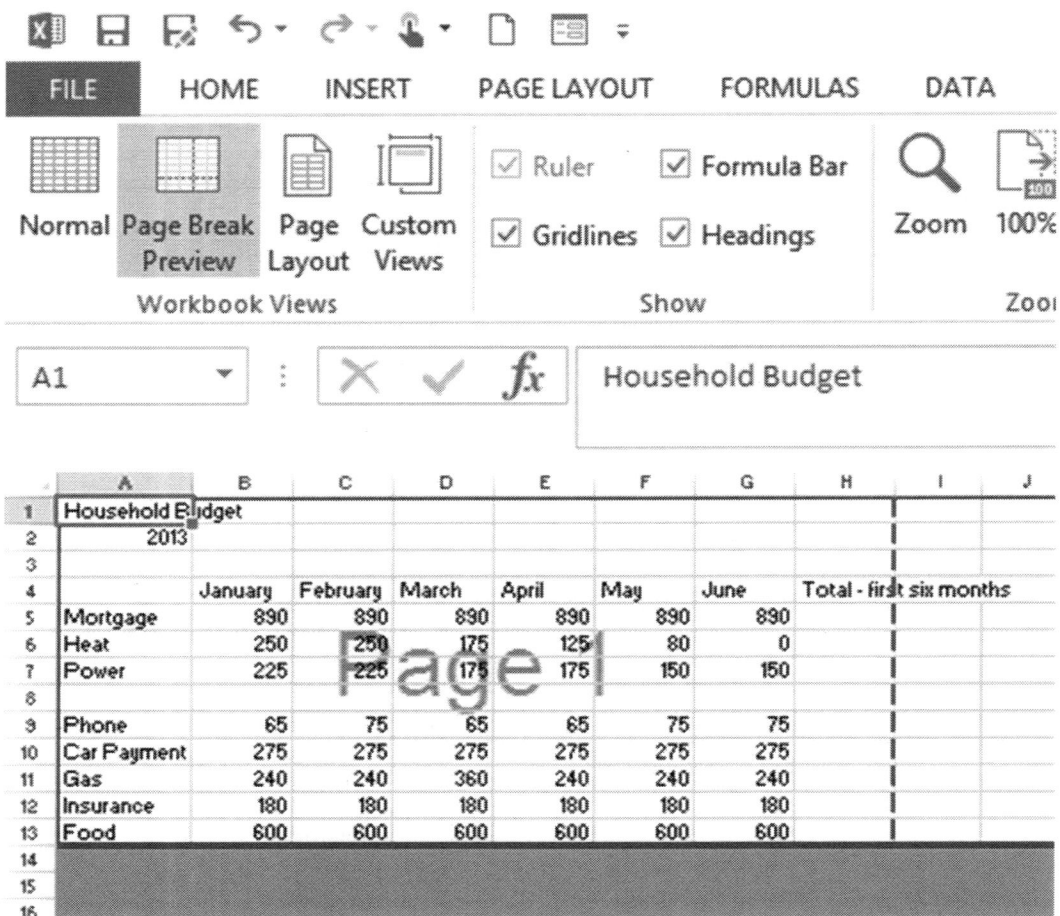

Investigate what happens if you move the blue borders.

Explore the Page Layout View, use the following procedure.

Step 1: Select the View tab.

Step 2: Select the Page Layout tool.

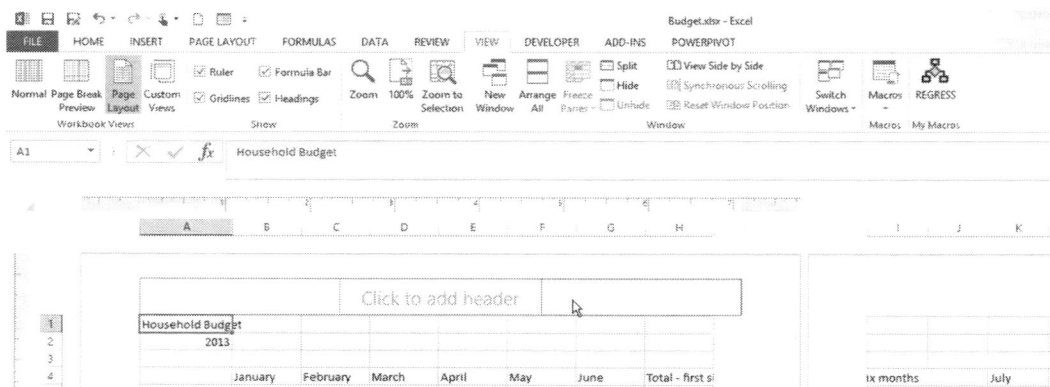

There are three areas for the header and the footer. Practice entering header content in the left, middle, and right of the header and/or footer.

Using Zoom

To zoom to a selection, use the following procedure.

Step 1: Highlight the area you want to view larger.

Step 2: Select the Zoom to Selection tool from the View tab on the Ribbon.

Step 3: Select 100% from the View tab on the Ribbon to return to the default zoom.

Creating Custom Views

To create a custom view, use the following procedure.

Step 1: Select Custom Views from the View tab on the Ribbon.

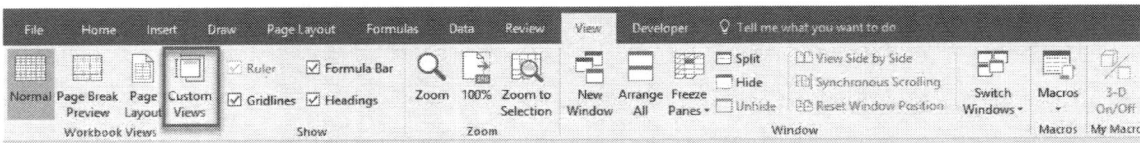

Excel opens the Custom Views dialog box.

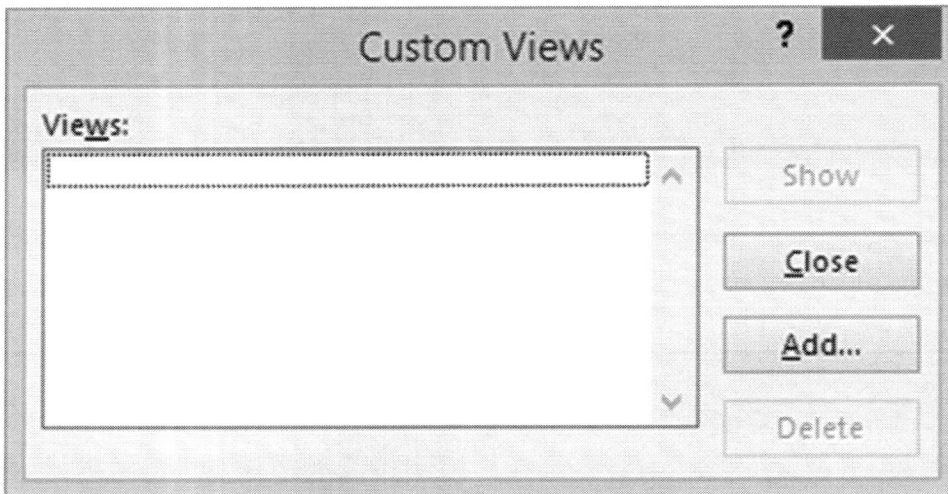

Step 2: Select Add to open the Add View dialog box.

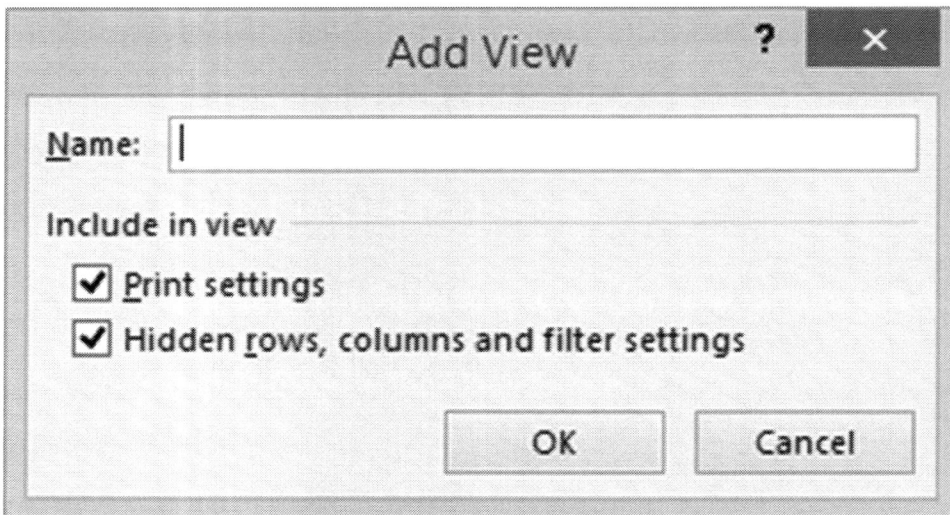

Step 3: Enter the Name of your view.

Step 4: Check the Print Settings box to include the print settings in your custom view.

Step 5: Check the Hidden rows, columns and filter settings to include those in your custom view.

Step 6: Select OK.

To apply a custom view, use the following procedure.

Step 1: Select Custom Views from the View tab on the Ribbon.

Step 2: Highlight the View you want to apply and select Show.

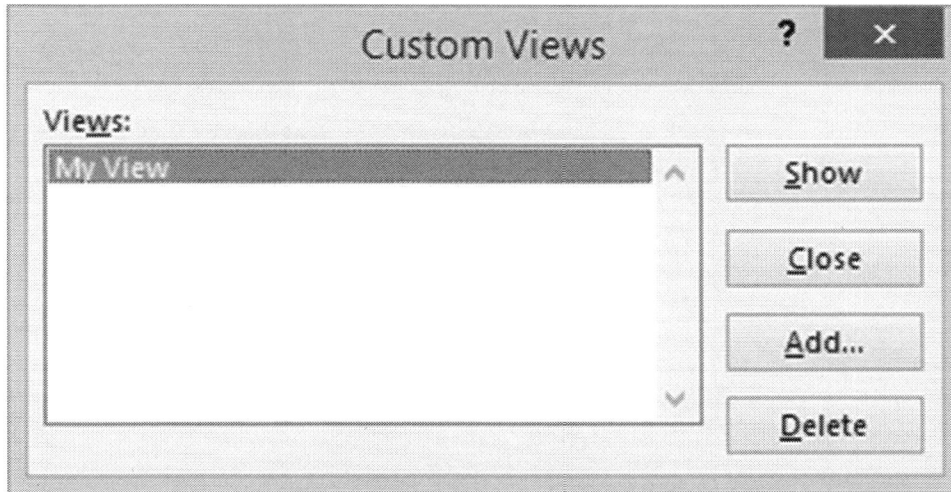

Switching Between Open Files

To switch from one worksheet to another, use the following procedure.

Step 1: Select the Switch Windows tool from the View tab on the Ribbon. Select the worksheet you want to view from the list.

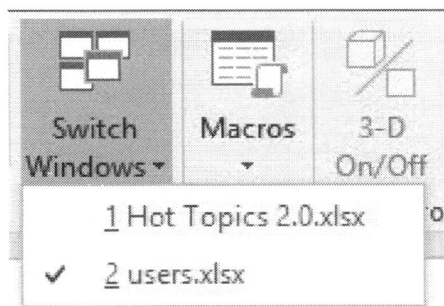

The backbone of Excel is its ability to perform calculations. There are two ways to set up calculations in Excel: using formulas or using functions. Formulas are mathematical expressions that you build yourself. You need to follow proper math principles to obtain the expected answer. Building the formula is simply a matter of combining the proper cell addresses with the correct operators in the right order. This chapter will explore how to build, edit, and copy formulas. This chapter will also explain the difference between relative and absolute references. Finally, this chapter will explain how to use the Status Bar to perform simple calculations. We will explore functions in the next chapter.

The Math Basics of Excel

Review the different types of operators.

The Arithmetic operators are:

- Plus Sign (+) – Adds values
- Minus Sign (-) – Subtracts values
- Asterisk (*) – Multiplies values
- Forward slash (/) – Divides values
- Percent sign (%) – Finds the percentage of a value
- Caret (^) – Exponentiation – Finds the exponential value The Comparison operators are:

- Equals (=) sign – Equates values
- Greater than (>) sign – Indicates that one value is greater than the other
- Less than sign (<) – Indicates that one value is less than the other
- Greater than or equal to (>=) signs – Indicates that one value is greater than or equal to the other
- Less than or equal to (<=) signs – Indicates that one value is less than or equal to the other

- Not Equal (<>) – Indicates that values are not equal

Text concatenation allows you to combine text from different cells into a single piece of text. The operator is the & sign.

The reference operators combine a range of cells to use together in an operation. The reference operators are:

- Colon (:) – A Range operator that produces a reference to all the cells between the references on either side of the colon
- Comma (,) – A Union operator that combines multiple range references
- Space – An intersection operator that returns a reference to the cells common to the ranges in the formula

Building a Formula

To enter a formula to calculate the Total Value in the sample worksheet, use the following procedure.

Step 1: Select the Total Value column for the first product (cell D4).

Step 2: Enter the = sign to begin the formula.

Step 3: Select cell B4 to use it as the first value in the formula. Excel enters the reference as part of the formula.

Cell reference: B4 — formula bar shows `=B4`

	A	B	C	D	E	F
1			Inventory			
2						
3	Part No.	# In Stock	Unit Price	Total Value	Reorder level	# left to reorder
4	QS12578	26	$ 248.89	=B4	20	
5	DSP4543	14	$ 124.50		10	
6	DS45848	2	$ 588.00		1	
7	SS12566	18	$ 224.67		10	
8	SSP2777	12	$ 118.00		5	
9	QS12585	5	$ 555.22		5	
10	DS12566	2	$ 470.99		1	
11	DS12556	8	$ 430.37		5	
12	KSP4333	4	$ 585.00		2	
13	QP133	12	$ 255.23		10	
14	KS36678	3	$ 685.75		1	
15						
16	Tax rate	10%				
17						

Step 4: Enter the * sign.

Step 5: Click on cell C4 to use it as the second value in the formula. Excel enters the references as part of the formula.

Clipboard   Font   Alignment

C4   =B4*C4

	A	B	C	D	E	F	G	H
1				Inventory				
2								
3	Part No.	# In Stock	Unit Price	Total Value	Reorder level	# left to reorder		
4	QS12578	26	$248.89	=B4*C4	20			
5	DSP4543	14	$124.50		10			
6	DS45848	2	$588.00		1			
7	SS12566	18	$224.67		10			
8	SSP2777	12	$118.00		5			
9	QS12585	5	$555.22		5			
10	DS12566	2	$470.99		1			
11	DS12556	8	$430.37		5			
12	KSP4333	4	$585.00		2			
13	QP133	12	$255.23		10			
14	KS36678	3	$685.75		1			
15								
16	Tax rate	10%						
17								

Step 6: Press ENTER to complete the formula. Excel moves to the next row and performs the calculations in the formula.

The following illustration shows the answer to the calculation in the cell, and since the cell is active, you can see the formula in the Formula bar.

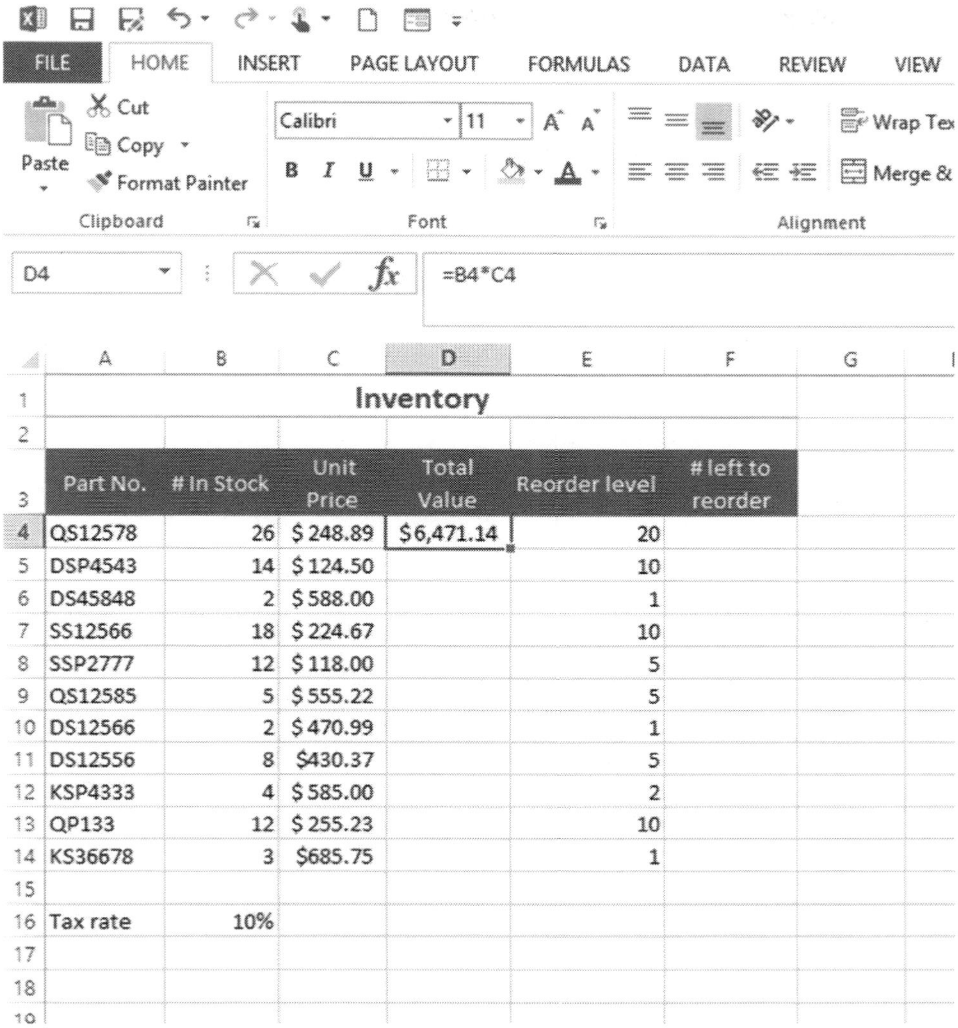

Editing a Formula

To edit a formula, use the following procedure. The following example uses an incorrect cell reference in a formula.

Step 1: Select the cell with the formula you want to correct to make it active.

Step 2: Select the Formula Bar. Excel highlights the cell references in the current formula.

⊿	A	B	C	D	E	F	G	H
1				Inventory				
2								
3	Part No.	# In Stock	Unit Price	Total Value	Reorder level	# left to reorder		
4	QS12578	26	$ 248.89	=B4*C4	20			
5	DSP4543	14	$ 124.50		10			
6	DS45848	2	$ 588.00		1			
7	SS12566	18	$ 224.67		10			
8	SSP2777	12	$ 118.00		5			
9	QS12585	5	$ 555.22		5			
10	DS12566	2	$ 470.99		1			
11	DS12556	8	$430.37		5			
12	KSP4333	4	$ 585.00		2			
13	QP133	12	$ 255.23		10			
14	KS36678	3	$685.75		1			
15								
16	Tax rate	10%						
17								
18								
19								
20								

Step 3: Highlight the operator or cell references and either type over with the correct reference or operator, or select the correct cell to replace a cell reference.

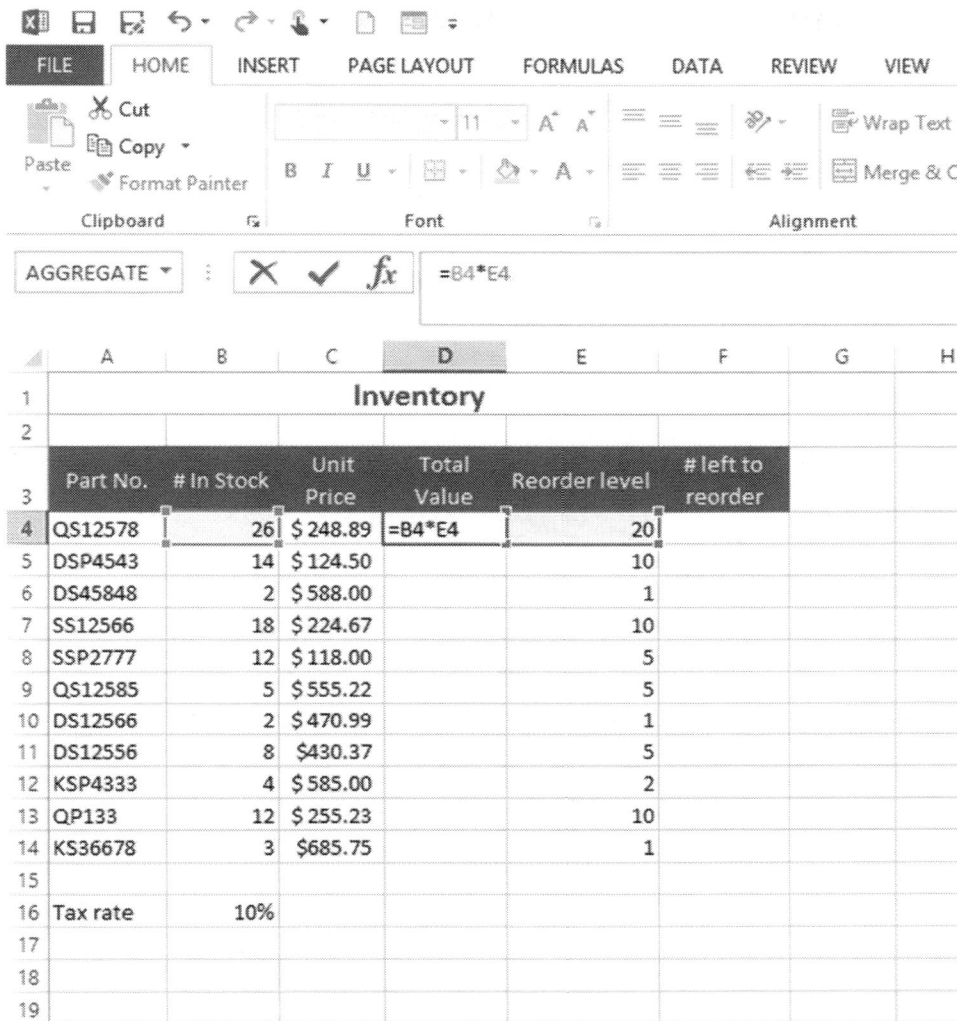

	A	B	C	D	E	F	G	H
1				Inventory				
2								
3	Part No.	# In Stock	Unit Price	Total Value	Reorder level	# left to reorder		
4	QS12578	26	$ 248.89	=B4*E4	20			
5	DSP4543	14	$ 124.50		10			
6	DS45848	2	$ 588.00		1			
7	SS12566	18	$ 224.67		10			
8	SSP2777	12	$ 118.00		5			
9	QS12585	5	$ 555.22		5			
10	DS12566	2	$ 470.99		1			
11	DS12556	8	$430.37		5			
12	KSP4333	4	$ 585.00		2			
13	QP133	12	$ 255.23		10			
14	KS36678	3	$685.75		1			
15								
16	Tax rate	10%						
17								
18								
19								

Step 4: Press ENTER to complete the formula. Excel calculates the formula and moves to the next row.

Copying a Formula

To copy and paste a formula, use the following procedure.

Step 1: Select the cell with the formula you want to copy. You can also click on the cell and use the keyboard shortcut: CTRL + C.

Step 2: Select Copy from the Home tab on the Ribbon.

	A	B	C	D	E	F	G	H
1				Inventory				
2								
3	Part No.	# In Stock	Unit Price	Total Value	Reorder level	# left to reorder		
4	QS12578	26	$ 248.89	$6,471.14	20	6		
5	DSP4543	14	$ 124.50		10			
6	DS45848	2	$ 588.00		1			
7	SS12566	18	$ 224.67		10			
8	SSP2777	12	$ 118.00		5			
9	QS12585	5	$ 555.22		5			
10	DS12566	2	$ 470.99		1			
11	DS12556	8	$430.37		5			
12	KSP4333	4	$ 585.00		2			
13	QP133	12	$ 255.23		10			
14	KS36678	3	$685.75		1			
15								
16	Tax rate	10%						
17								
18								
19								

Excel highlights the cell whose contents you are copying. This will remain highlighted until you finish pasting, in case you want to paste the cell contents more than once.

Select the cell where you want to copy the formula. Excel displays several paste options. To paste a formula, select Paste or Paste formula. Note that as you hover your mouse over the paste options, the rest of the context menu is dimmed. You can also select the cell and use the keyboard shortcut: CTRL + V.

	A	B	C	D	E	F	G	H	I
1				Inventory					
2									
3	Part No.	# In Stock	Unit Price	Total Value	Reorder level	# le reo			
4	QS12578	26	$248.89	$6,471.14	20				
5	DSP4543	14	$124.50		10				
6	DS45848	2	$588.00		1				
7	SS12566	18	$224.67		10				
8	SSP2777	12	$118.00		5				
9	QS12585	5	$555.22		5				
10	DS12566	2	$470.99		1				
11	DS12556	8	$430.37		5				
12	KSP4333	4	$585.00		2				
13	QP133	12	$255.23		10				
14	KS36678	3	$685.75		1				

Step 1: You can repeat the paste as many times as desired. Or you can highlight multiple cells at once before pasting to repeat the paste for all highlighted cells.

Step 2: Press ENTER to stop pasting.

Relative versus Absolute References

To copy a formula with an absolute reference, use the following procedure.

Step 1: Create a new column labeled Taxes.

Step 2: Select the Taxes column for the first product (cell E4).

Step 3: Enter the = sign to begin the formula.

Step 4: Select cell B16 to use it as the first value in the formula. Excel enters the reference as part of the formula. Use the Formula Bar to enter dollar signs before the column and the row (i.e., $B$16).

Step 5: Enter * and the relative reference in the Total Value column.

	A	B	C	D	E	F	G	
1				Inventory				
2								
3	Part No.	# In Stock	Unit Price	Total Value	Taxes	Reorder level	# left to reorder	
4	QS12578	26	$ 248.89	$6,471.14	=$B$16*D4	20	6	
5	DSP4543	14	$ 124.50			10		
6	DS45848	2	$ 588.00			1		
7	SS12566	18	$ 224.67			10		
8	SSP2777	12	$ 118.00			5		
9	QS12585	5	$ 555.22			5		
10	DS12566	2	$ 470.99			1		
11	DS12556	8	$430.37			5		
12	KSP4333	4	$ 585.00			2		
13	QP133	12	$ 255.23			10		
14	KS36678	3	$685.75			1		
15								
16	Tax rate	10%						
17								
18								

Step 6: Press ENTER to complete the formula. Excel moves to the next row and performs the calculations in the formula.

Copy the formula for the other products and select some of them to see the results.

Using the Status Bar to Perform Calculations

To customize the Status Bar, use the following procedure.

Step 1: Right click on the Status Bar to see a list of Functions that can be displayed. For this example, select MIN and MAX.

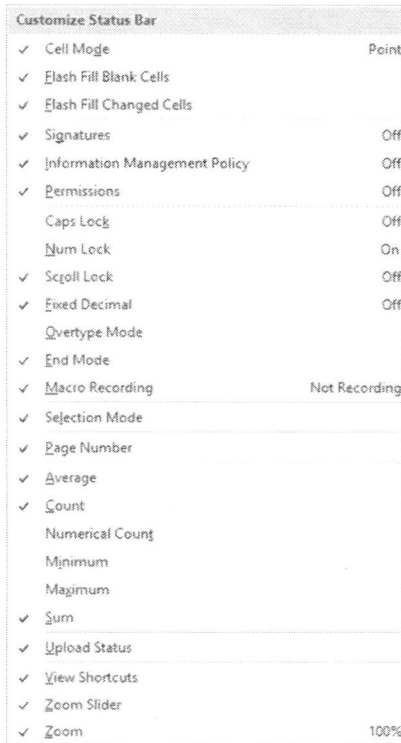

Customize Status Bar	
✓ Cell Mode	Point
✓ Flash Fill Blank Cells	
✓ Flash Fill Changed Cells	
✓ Signatures	Off
✓ Information Management Policy	Off
✓ Permissions	Off
Caps Lock	Off
Num Lock	On
✓ Scroll Lock	Off
✓ Fixed Decimal	Off
Overtype Mode	
✓ End Mode	
✓ Macro Recording	Not Recording
✓ Selection Mode	
✓ Page Number	
✓ Average	
✓ Count	
Numerical Count	
Minimum	
Maximum	
✓ Sum	
✓ Upload Status	
✓ View Shortcuts	
✓ Zoom Slider	
✓ Zoom	100%

Step 2: Press ESC or click elsewhere in the worksheet to close the Customize Status Bar list.

Explore the calculations performed when you highlight a group of cells.

	A	B	C	D	E	F	G	H	I
4		January	February	March	April	May	June	Total - first six months	
5	Mortgage	890	890	890	890	890	890		
6	Heat	250	250	175	125	80	0		
7	Power	225	225	175	175	150	150		
8									
9	Phone	65	75	65	65	75	75		
10	Car Payment	275	275	275	275	275	275		
11	Gas	240	240	360	240	240	240		
12	Insurance	180	180	180	180	180	180		
13	Food	600	600	600	600	600	600		
14									
15									
16									

Chapter 6 – Using Excel Functions

This chapter introduces Excel functions, which are a little like templates for common formulas. There are many different types of functions. First, we will look at the SUM function. You will learn about using AutoComplete for entering formulas. We will look at other basic common functions as well. We will look at the Formulas tab introduced in the Ribbon for Excel 2007. Finally, we will look at the function names.

Formulas versus Functions

To open the Insert Function dialog box, use the following procedure.

Step 1: Select the Insert Function tool right next to the Formula Bar.

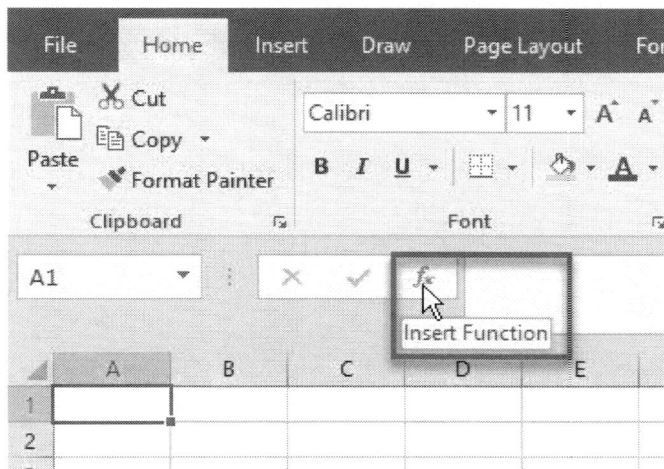

Investigate the different categories and functions in the Insert Function dialog box. Point out the bottom part of the screen where the syntax and description of the function appear.

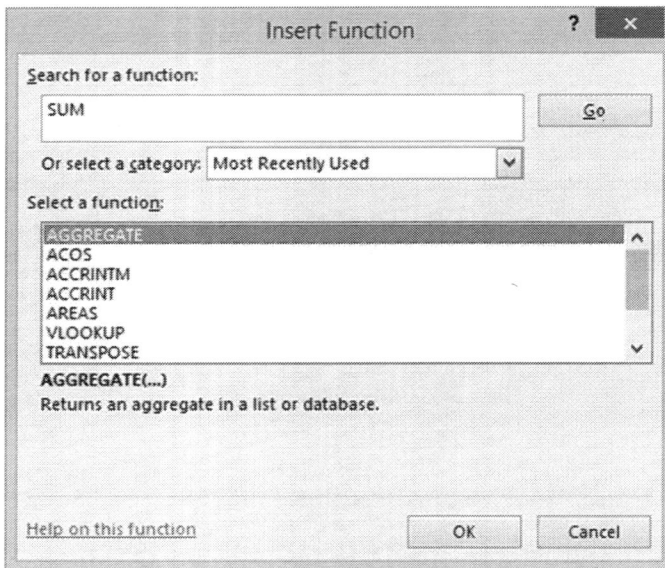

Using the SUM Function

Review how to use a SUM function to add the total for each category in the sample file, use the following procedure.

Step 1: Select the Total –First Six Months column for the first category (cell H5).

Step 2: Select the AutoSum tool in the Editing Group on the Home tab of the Ribbon.

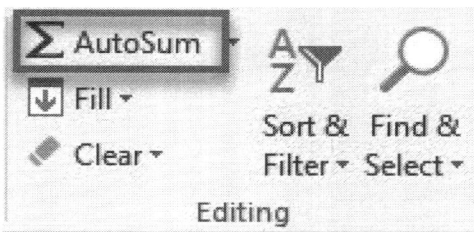

Step 3: Excel enters the function with a default selection of the cell references you want to use in the function highlighted.

The formula bar shows: `=SUM(B5:G5)`

	A	B	C	D	E	F	G	H	I	J
1	Household Budget									
2	2013									
3										
4		January	February	March	April	May	June	Total - first six months		
5	Mortgage	890	890	890	890	890	890	=SUM(		
6	Heat	250	250	175	125	80	0	SUM(number1, [number2], ...)		
7	Power	225	225	175	175	150	150			
8										
9	Phone	65	75	65	65	75	75			
10	Car Payment	275	275	275	275	275	275			
11	Gas	240	240	360	240	240	240			
12	Insurance	180	180	180	180	180	180			
13	Food	600	600	600	600	600	600			
14										
15										
16										
17										

Step 4: If the cell references are not accurate, you can drag the highlighted area to include additional cells or remove cells you do not want used in the function.

Step 5: Press ENTER to complete the function.

Excel performs the calculation and moves to the next row. In the following illustration, the cell with the function is active, so that you can see the function syntax in the Formula Bar and the result in the cell.

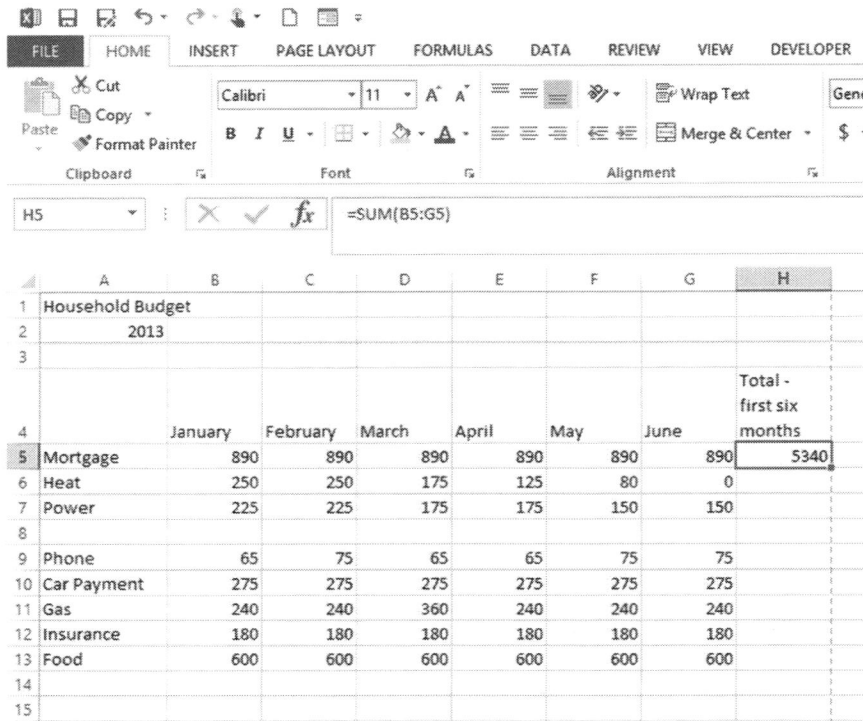

	A	B	C	D	E	F	G	H
1	Household Budget							
2	2013							
3								
4		January	February	March	April	May	June	Total - first six months
5	Mortgage	890	890	890	890	890	890	5340
6	Heat	250	250	175	125	80	0	
7	Power	225	225	175	175	150	150	
8								
9	Phone	65	75	65	65	75	75	
10	Car Payment	275	275	275	275	275	275	
11	Gas	240	240	360	240	240	240	
12	Insurance	180	180	180	180	180	180	
13	Food	600	600	600	600	600	600	
14								
15								

Using AutoComplete

To use the AutoComplete feature, use the following procedure.

Step 1: Begin typing the SUM function. As soon as you type the Equals sign and the letter S, Excel displays a possible list of matching functions.

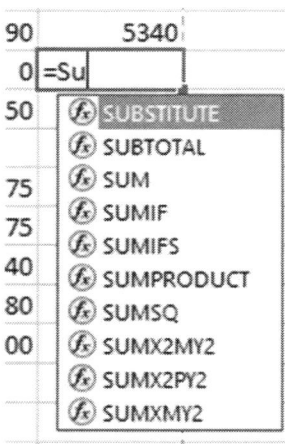

Step 2: To select the SUM Function from the list, double-click on the SUM function.

Step 3: Excel enters the function, but you must still enter the arguments. You can simply click on multiple cells, or click and drag to select a cell range. You can also type in the cell references.

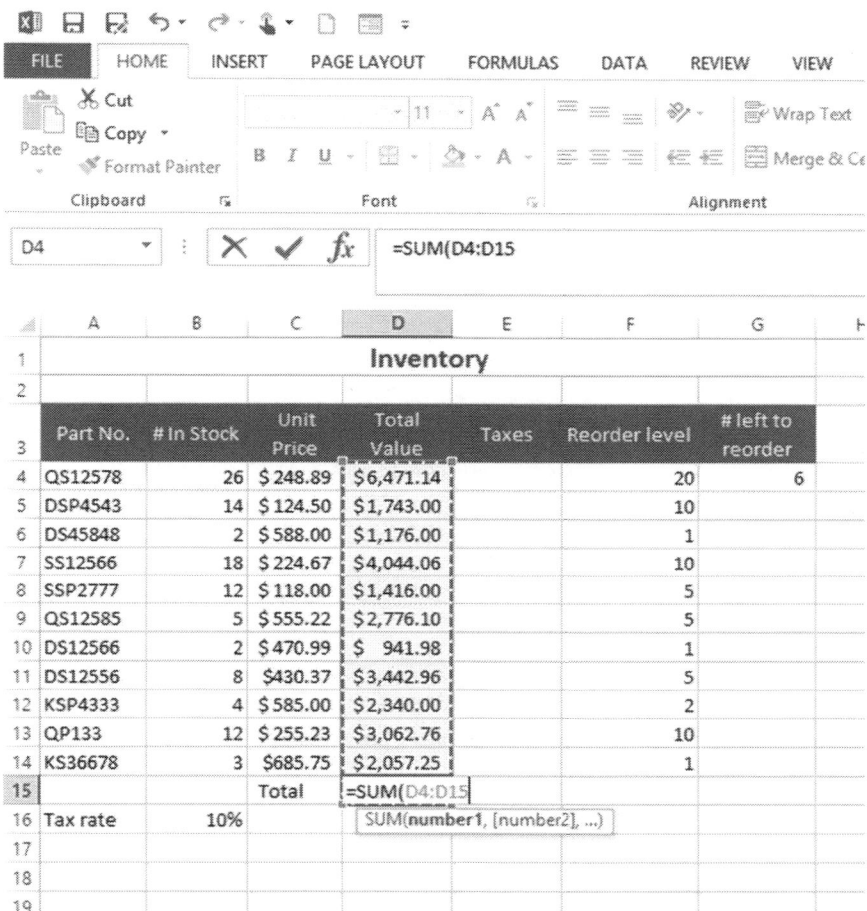

| | Cell: D4 | | | | fx | =SUM(D4:D15 | | | |

	A	B	C	D	E	F	G	H
1				Inventory				
2								
3	Part No.	# In Stock	Unit Price	Total Value	Taxes	Reorder level	# left to reorder	
4	QS12578	26	$248.89	$6,471.14		20	6	
5	DSP4543	14	$124.50	$1,743.00		10		
6	DS45848	2	$588.00	$1,176.00		1		
7	SS12566	18	$224.67	$4,044.06		10		
8	SSP2777	12	$118.00	$1,416.00		5		
9	QS12585	5	$555.22	$2,776.10		5		
10	DS12566	2	$470.99	$ 941.98		1		
11	DS12556	8	$430.37	$3,442.96		5		
12	KSP4333	4	$585.00	$2,340.00		2		
13	QP133	12	$255.23	$3,062.76		10		
14	KS36678	3	$685.75	$2,057.25		1		
15			Total	=SUM(D4:D15				
16	Tax rate	10%		SUM(number1, [number2], ...)				
17								
18								
19								

Step 4: Enter the final parenthesis mark to end the function.

Step 5: Press ENTER to enter the function in the cell.

Using Other Basic Excel Functions

Review how to use the AVERAGE function as an example of another function, use the following procedure.

Step 1: Add a new label in column I: Average.

Step 2: Select the cell in the Average column for the first category.

Step 3: Select the arrow next to the SUM function on the Home tab of the Ribbon to see the list of other common functions.

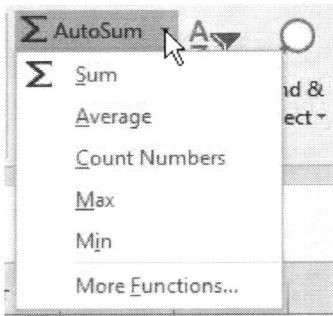

Step 4: Select Average.

Excel enters the function with the most likely cell references.

4		January	February	March	April	May	June	Total - first six months	Average
5	Mortgage	890	890	890	890	890	890	5340	=AVERAGE(
6	Heat	250	250	175	125	80	0		AVERAGE(number1, [number2], ...)
7	Power	225	225	175	175	150	150		
8									
9	Phone	65	75	65	65	75	75		

Step 5: Replace the cell references so that cell H5 is not included in the average.

4		January	February	March	April	May	June	Total - first six months	Average
5	Mortgage	890	890	890	890	890	890	5340	=AVERAGE(B5:G5)
6	Heat	250	250	175	125	80	0		AVERAGE(number1, [number2], ...)
7	Power	225	225	175	175	150	150		
8									
9	Phone	65	75	65	65	75	75		

Step 6: Press ENTER to complete the function.

Understanding the Formulas Tab

Explore the Formulas tab on the Ribbon.

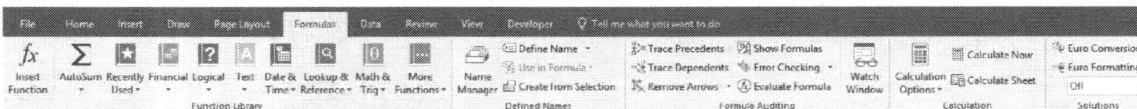

Explore the icons that appear when entering a function name. The old names have a yellow warning triangle next to them.

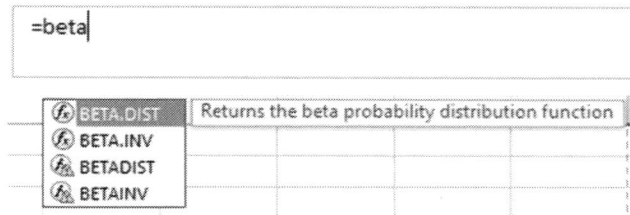

```
=beta
```

fx BETA.DIST	Returns the beta probability distribution function
fx BETA.INV	
BETADIST	
BETAINV	

Chapter 7 – Using Quick Analysis

The new Quick Analysis tools allow you to easily preview and apply many formatting, charts, totals, tables, and spark lines features to your data. We will first look at the formatting tools. You will also learn about the recommended charts available in Quick Analysis. Next, we will look at the totals tools that include formulas that automatically calculate totals for you. We will also look at the tables available with Quick Analysis. Once you have your data in table format, you can use several sorting and filtering tools, which we will also discuss in this chapter. Finally, you will learn how to create spark lines.

Formatting Data

To apply Quick Analysis formatting, use the following procedure.

Step 1: Select the table (A1 to D16) in the sample worksheet.

Step 2: Select the icon that appears at the bottom right of the table.

		Number	Percent	Number	Percent	Number	Percent
7							
8	Individual	297,229	235,721	79.3	222,038	74.7	205,500	69.1
9								
10	.Age						
11	..3-17 ye	62,002	50,835	82	42,843	69.1	38,461	62
12	..18-34 y	71,155	58,624	82.4	61,174	86	55,966	78.7
13	..35-44 y	39,629	33,459	84.4	33,292	84	31,280	78.9
14	..45-64 y	82,066	66,489	81	62,962	76.7	59,177	72.1
15	..65 yeal	42,377	26,315	62.1	21,767	51.4	20,616	48.7

Quick Analysis (Ctrl+Q)

Use the Quick Analysis tool to quickly and easily analyze your data with some of Excel's most useful tools, such as charts, color-coding, and formulas.

Step 3: Select the formatting type that you want to use.

7			Number	Percent	Number	Percent	Number	Percent
8	Individual	297,229	235,721	79.3	222,038	74.7	205,500	69.1
9								
10	.Age							
11	..3-17 ye	62,002	50,835	82	42,843	69.1	38,461	62
12	..18-34 y	71,155	58,624	82.4	61,174	86	55,966	78.7
13	..35-44 y	39,629	33,459	84.4	33,292	84	31,280	78.9
14	..45-64 y	82,066	66,489	81	62,962	76.7	59,177	72.1
15	..65 year	42,377	26,315	62.1	21,767	51.4	20,616	48.7

	Formatting	Charts	Totals	Tables	Sparklines
	Data Bars	Color...	Icon Set	Greater...	Top 10%	Clear...

Conditional Formatting uses rules to highlight interesting data.

Step 4: For the Greater Than option (and some other types of options), enter the cell that contains the value to which you want to compare the others. You can also enter a number or a formula. Also select the formatting you want to use from the drop-down list.

	A	B	C	D
1	Company	Industry	Q1 Sales	Q2 Sales
2	A. Datum Corporation	Tech	$195,449	$746,907
3	Adventure Works	Travel	$123,721	$733,396
4	Blue Yonder Airlines	Travel	$934,763	$246,554
5	City Power & Light	Utilities	$299,293	$674,295
6	Coho Vineyard	Beverage	$228,783	$659,385
7	Contoso, Ltd	Misc	$239,219	$287,989
8	Contoso Pharmaceuticals	Medical	$371,570	$644,368
9	Consolidated Messenger	Tech	$579,825	$448,399
10	Fabrikam, Inc.	Utilities	$639,630	$635,474
11	Fourth Coffee	Beverage	$876,740	$567,216
12	Graphic Design Institute	Education		
13	Humongous Insurance	Financial		
14	Litware, Inc.	Tech		
15	Lucerne Publishing	Misc		
16	Margie's Travel	Travel		

Greater Than

Format cells that are GREATER THAN:

$529,242   with   Light Red Fill with Dark Red Text

OK   Cancel

To remove any formatting you have applied, select Clear format.

Creating Quick Analysis Charts

To create Quick Analysis charts, use the following procedure.

Step 1: Select the table (A1 to D16) in the sample worksheet.

Step 2: Select the icon that appears at the bottom right of the table.

Step 3: Select Charts.

Step 4: Select the chart type that you want to use.

7			Number	Percent	Number	Percent	Number	Percent	
8	Individual	297,229	235,721	79.3	222,038	74.7	205,500	69.1	
9									
10	.Age								
11	..3-17 ye	62,002	50,835	82	42,843	69.1	38,461	62	
12	..18-34 y	71,155	58,624	82.4	61,174	86	55,966	78.7	
13	..35-44 y	39,629	33,459	84.4	33,292	84	31,280	78.9	
14	..45-64 y	82,066	66,489	81	62,962	76.7	59,177	72.1	
15	..65 yea|	42,377	26,315	62.1	21,767	51.4	20,616	48.7	
16									
17									

Formatting   Charts   Totals   Tables   Sparklines

Clustere...   More...

Recommended Charts help you visualize data.

The chart is inserted into your worksheet. If the chart you selected is a PivotChart, it will be created on a separate worksheet.

Calculating Totals

To create Quick Analysis totals, use the following procedure.

Step 1: Select the table (A1 to D16) in the sample worksheet.

Step 2: Select the icon that appears at the bottom right of the table.

Step 3: Select Totals.

Step 4: Select the formula that you want to use. Note that there is a right arrow to view additional options.

7			Number	Percent	Number	Percent	Number	Percent	
8	Individual	297,229	235,721	79.3	222,038	74.7	205,500	69.1	
9									
10	.Age								
11	..3-17 ye	62,002	50,835	82	42,843	69.1	38,461	62	
12	..18-34 y	71,155	58,624	82.4	61,174	86	55,966	78.7	
13	..35-44 y	39,629	33,459	84.4	33,292	84	31,280	78.9	
14	..45-64 y	82,066	66,489	81	62,962	76.7	59,177	72.1	
15	..65 yea|	42,377	26,315	62.1	21,767	51.4	20,616	48.7	
16									
17									

Formatting   Charts   Totals   Tables   Sparklines

Sum   Average   Count   % Total   Running...   Sum

Formulas automatically calculate totals for you.

The row or column you selected is inserted into your worksheet.

Creating Quick Analysis Tables

To create Quick Analysis tables, use the following procedure.

Step 1: Select the table (A1 to D16) in the sample worksheet.

Step 2: Select the icon that appears at the bottom right of the table.

Step 3: Select Tables.

Step 4: Select the table type that you want to use.

7			Number	Percent	Number	Percent	Number	Percent
8	Individual	297,229	235,721	79.3	222,038	74.7	205,500	69.1
9								
10	.Age							
11	..3-17 ye	62,002	50,835	82	42,843	69.1	38,461	62
12	..18-34 y	71,155	58,624	82.4	61,174	86	55,966	78.7
13	..35-44 y	39,629	33,459	84.4	33,292	84	31,280	78.9
14	..45-64 y	82,066	66,489	81	62,962	76.7	59,177	72.1
15	..65 year	42,377	26,315	62.1	21,767	51.4	20,616	48.7

Formatting | Charts | Totals | Tables | Sparklines

Table   Blank...

Tables help you sort, filter, and summarize data.

The Table is inserted into your worksheet with formatting and filtering options activated. If the table you selected is a PivotTable, it will be created on a separate worksheet.

Sorting and Filtering Your Data

To sort, use the following procedure.

Step 1: Select the arrow next to the column header that you want to use for sorting or filtering. Or you can select the Sort & Filter tool from the Ribbon.

	A	B	C	D	E
1	Company	Industry	Q1 Sales	Q2 Sales	
2	A. Datum Corporation		Sort Smallest to Largest		
3	Adventure Works		Sort Largest to Smallest		
4	Blue Yonder Airlines		Sort by Color		
5	City Power & Light				
6	Coho Vineyard		Clear Filter From "Q2 Sales"		
7	Contoso, Ltd		Filter by Color		
8	Contoso Pharmaceuticals		Number Filters		
9	Consolidated Messenger		Search		
10	Fabrikam, Inc.		☑ (Select All)		
11	Fourth Coffee		☑ $206,331		
12	Graphic Design Institute		☑ $246,554		
13	Humongous Insurance		☑ $287,989		
14	Litware, Inc.		☑ $443,552		
15	Lucerne Publishing		☑ $448,399		
16	Margie's Travel		☑ $540,282		
17			☑ $567,216		
18			☑ $577,599		
19			☑ $635,474		
20			OK Cancel		
21					
22					

Step 2: Select a sorting option from the list. The list differs, depending on the type of data in the column you selected.

To create a custom sort, use the following procedure.

Step 1: Select one column header you want to use in your sort.

Step 2: Select the Sort & Filter tool from the Ribbon.

Step 3: Select Custom Sort.

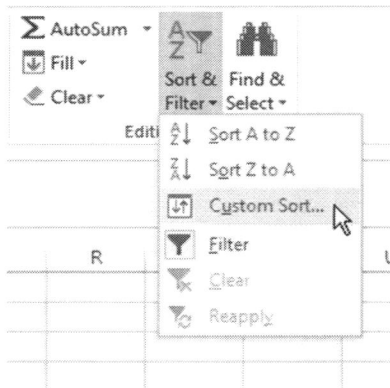

Excel opens the Sort dialog box.

Step 4: You can choose the first column by which to sort from the Sort By drop down list. The options displayed match the column headers in your worksheet.

Step 5: Select an option from the Sort On drop down list. Values is selected by default.

Step 6: Select an Order from the drop-down list.

Step 7: To add another column to your sort, select Add Level. Repeat steps 4, 5, and 6 for the next sorting level. You can Delete the Level, Copy a Level, and rearrange the order of the sorting levels by using the up or down arrows.

Step 8: Select OK when you have finished setting up your sort to see the results.

To apply a simple filter, use the following procedure.

Step 1: Select the arrow next to the column header for the column including the value you want to filter.

Step 2: To select a simple filter based on your type of data, select the option provided above the search box. In the example, below, select Number Filters. Then select the filter you want to use from the list. Excel will display a dialog box to determine the value to use in your filter.

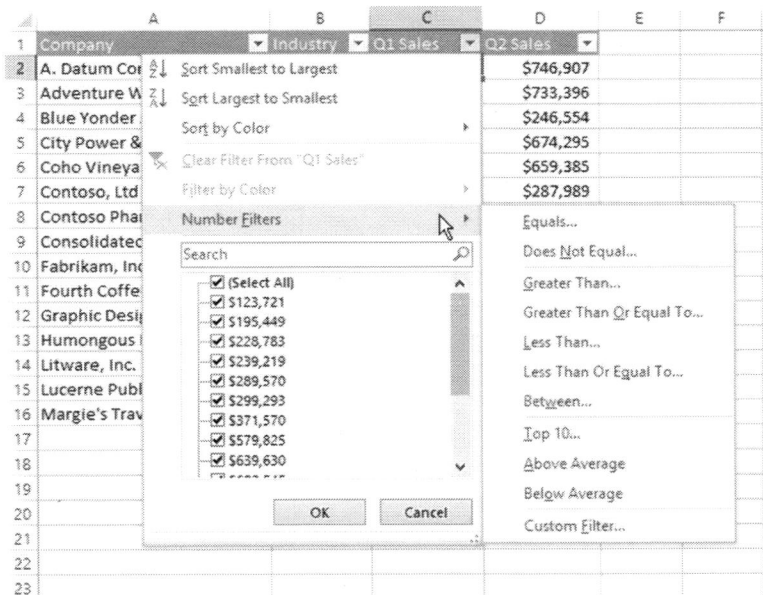

Step 3: Enter the value you want to use in the dialog box and select OK.

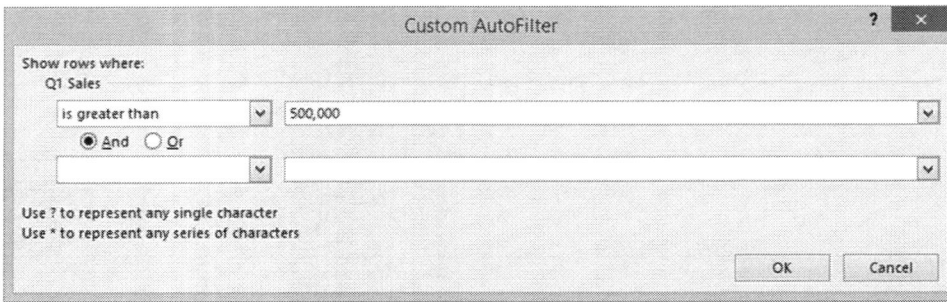

Custom AutoFilter

Show rows where:
Q1 Sales

| is greater than | ∨ | 500,000 | ∨ |

◉ And ○ Or

| | ∨ | | ∨ |

Use ? to represent any single character
Use * to represent any series of characters

| OK | Cancel |

Step 4: To select a simple filter based one or more of your specific values, clear the Select All check box to clear the boxes. Check one or more boxes from the items that are taken from your data. Select OK to apply the filter.

Excel includes a different small icon in the column header to indicate that a filter has been applied. It only shows the rows that match the filter. The other rows are still present, but hidden.

C	D
Q1 Sales	Q2 Sales
$934,763	$246,554
$579,825	$448,399
$639,630	$635,474
$876,740	$567,216
$788,390	$540,282
$682,545	$577,599
$902,264	$206,331
$905,906	$443,552

To clear a filter, use the following procedure.

Step 1: Select the Filter icon next to the column header to open the Sort and Filter context menu.

	A	B	C	D
	ny	Industry	Q1 Sales	Q2 Sales
	nder	Sort Smallest to Largest		$246,554
	idatec	Sort Largest to Smallest		$448,399
	m, Inc			$635,474
	Coffe	Sort by Color ▸		$567,216
	Desi	Clear Filter From "Q1 Sales"		$540,282
	gous	Filter by Color ▸		$577,599
	, Inc.	✓ Number Filters ▸		$206,331
	Publ			$443,552

Search

- ☐ (Select All)
- ☐ $123,721
- ☐ $195,449
- ☐ $228,783
- ☐ $239,219
- ☐ $289,570
- ☐ $299,293
- ☐ $371,570
- ☐ $579,825
- ☐ $639,630

OK   Cancel

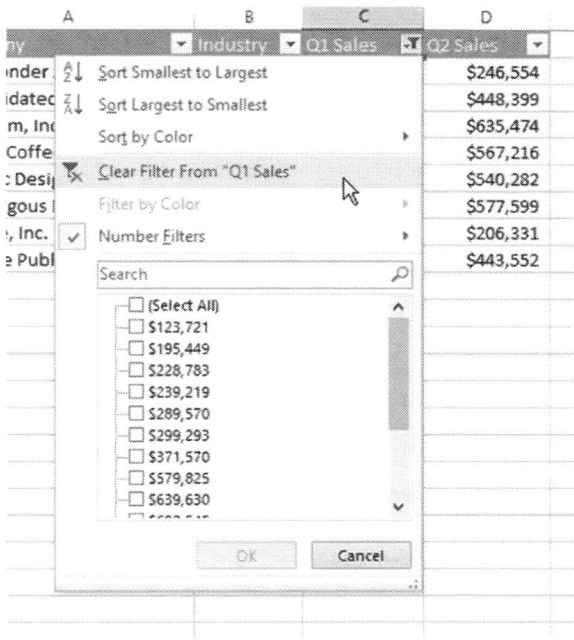

Step 2: Select Clear Filter From to clear the filter.

Using Spark Lines

To create Quick Analysis spark lines, use the following procedure.

Step 1: Select the table (A1 to D16) in the sample worksheet.

Step 2: Select the icon that appears at the bottom right of the table.

Step 3: Select Spark lines.

Step 4: Select the type of mini chart that you want to use.

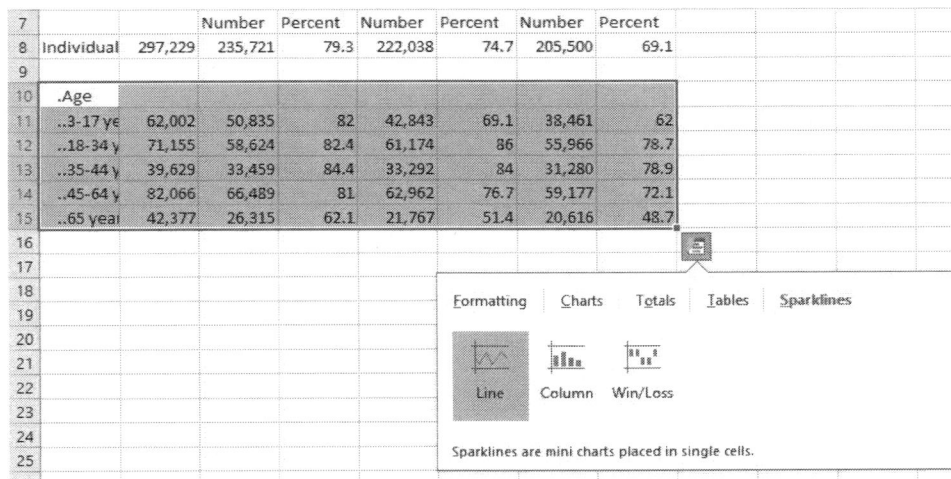

7			Number	Percent	Number	Percent	Number	Percent
8	Individual	297,229	235,721	79.3	222,038	74.7	205,500	69.1
9								
10	.Age							
11	..3-17 ye	62,002	50,835	82	42,843	69.1	38,461	62
12	..18-34 y	71,155	58,624	82.4	61,174	86	55,966	78.7
13	..35-44 y	39,629	33,459	84.4	33,292	84	31,280	78.9
14	..45-64 y	82,066	66,489	81	62,962	76.7	59,177	72.1
15	..65 yeai	42,377	26,315	62.1	21,767	51.4	20,616	48.7

Formatting   Charts   Totals   Tables   Sparklines

Line   Column   Win/Loss

Sparklines are mini charts placed in single cells.

Chapter 8 – Formatting Your Data

In this chapter, we will look at how to make your worksheet more appealing by changing the font type and size, alignment, formatting numbers, and by adding color and borders. This chapter also explains how to use the merge feature and how to remove formatting.

Changing the Appearance of Text

To apply formatting to text, use the following procedure.

Step 1: Select one or more cells that you want to format.

Step 2: Right-click to display the context menu, or use the formatting tools on the Home tab.

Step 2a: Use the Font drop down list to select a new font for the text.

Step 2b: Use the Font Size drop down list to select a new font size for the text. Alternatively, you can use the Increase Font Size or Decrease Font Size tools to adjust the font size a point at a time.

Step 2c: Select Bold, Italics, or Underline to add these features to your text.

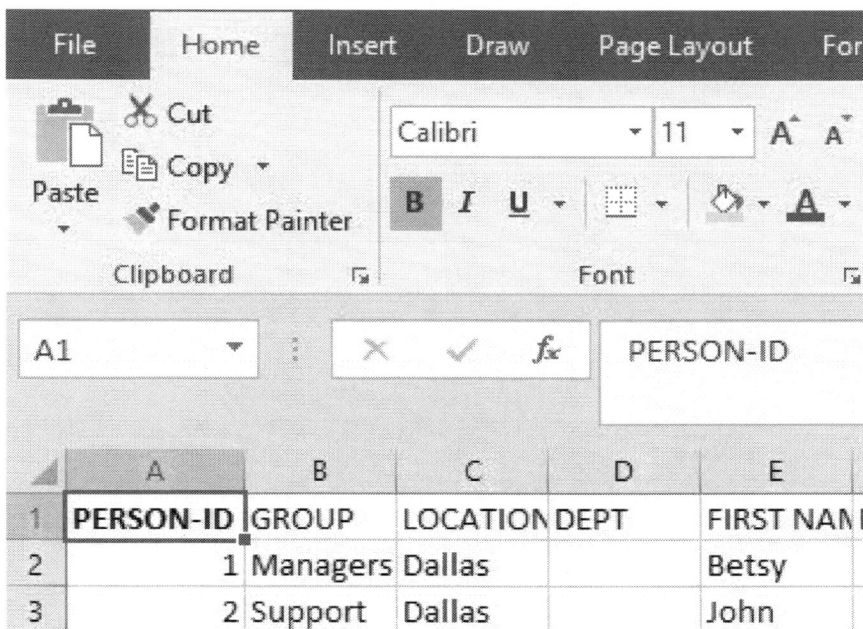

Changing the Appearance of Numbers

To format a number as currency without decimals, use the following procedure.

Step 1: Select the cell or cell range that you want to format.

Step 2: Select the type of number formatting you want to use from the Number group drop down list in the Home tab of the Ribbon.

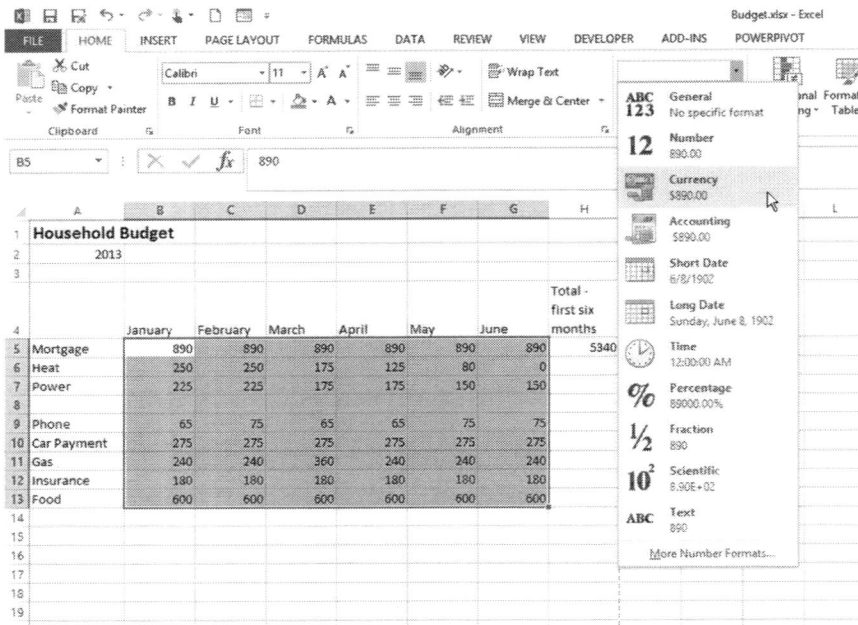

Step 3: Select the Decrease Decimal tool (2 times) to remove the decimal places.

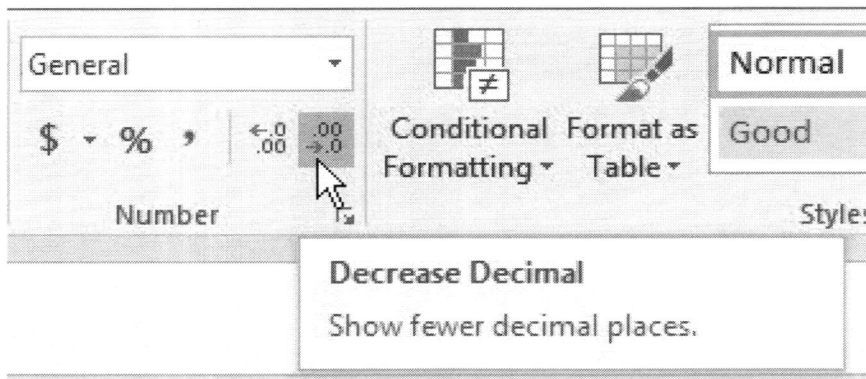

Adding Borders and Fill Color

To add borders, use the following procedure.

Step 1: Highlight the cell or cell range where you want to apply your border.

Step 2: Select the type of border you want to apply from the Borders tool on the Home tab of the Ribbon.

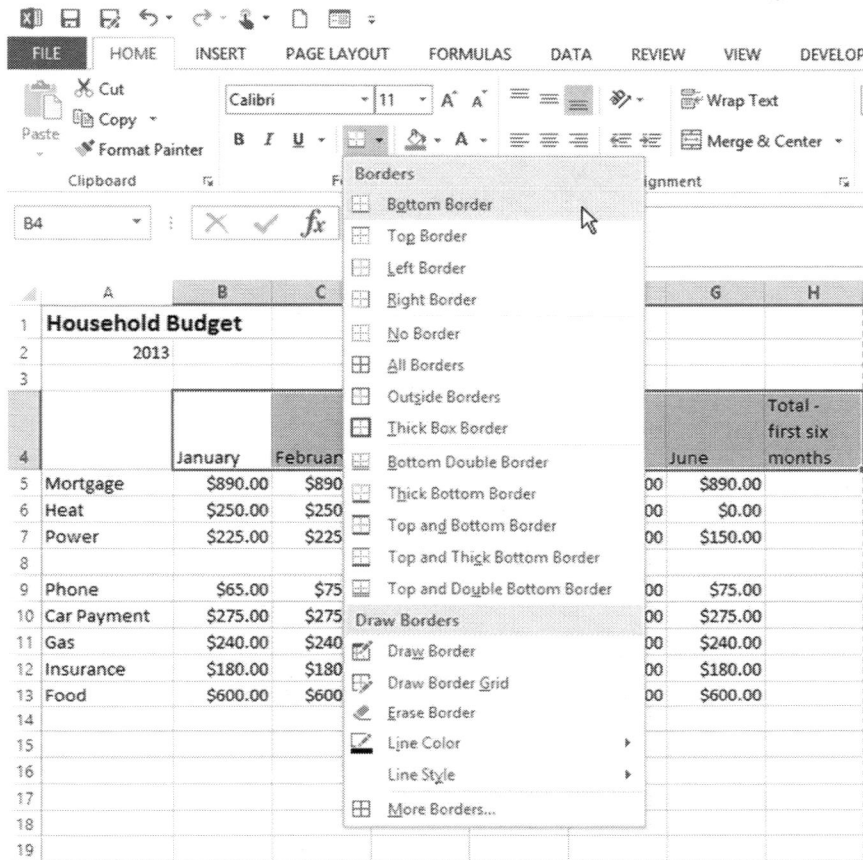

To apply fill colors, use the following procedure.

Step 1: Highlight the cell or cell range where you want to apply your fill color.

Step 2: Select the color you want to apply from the Fill Color tool on the Home tab of the Ribbon.

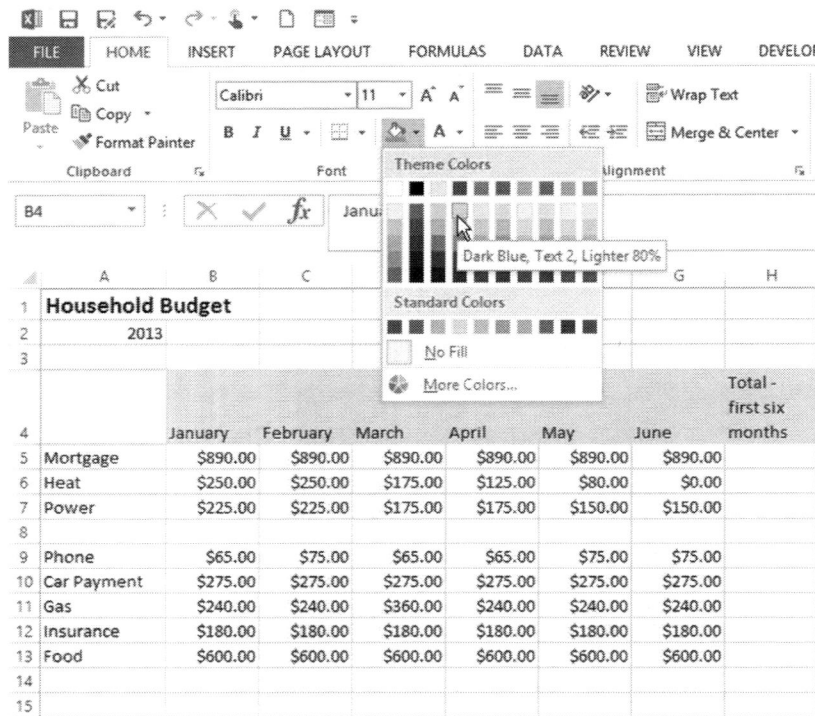

Removing Formatting

To clear formatting, use the following procedure.

Step 1: Select the cell or cell range that you want to clear.

Step 2: Select the Clear tool from the Home tab on the Ribbon.

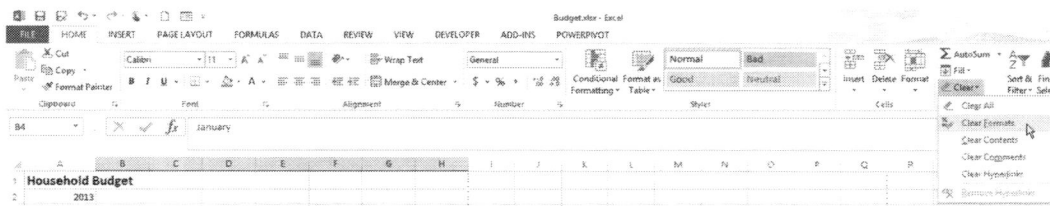

Step 3: Select Clear Formats.

Working with Alignment Options

To align cell contents, use the following procedure.

Step 1: Select the cell or cell range that you want to align.

Step 2: Select the type of alignment you want to use from the Alignment group tools in the Home tab of the Ribbon.

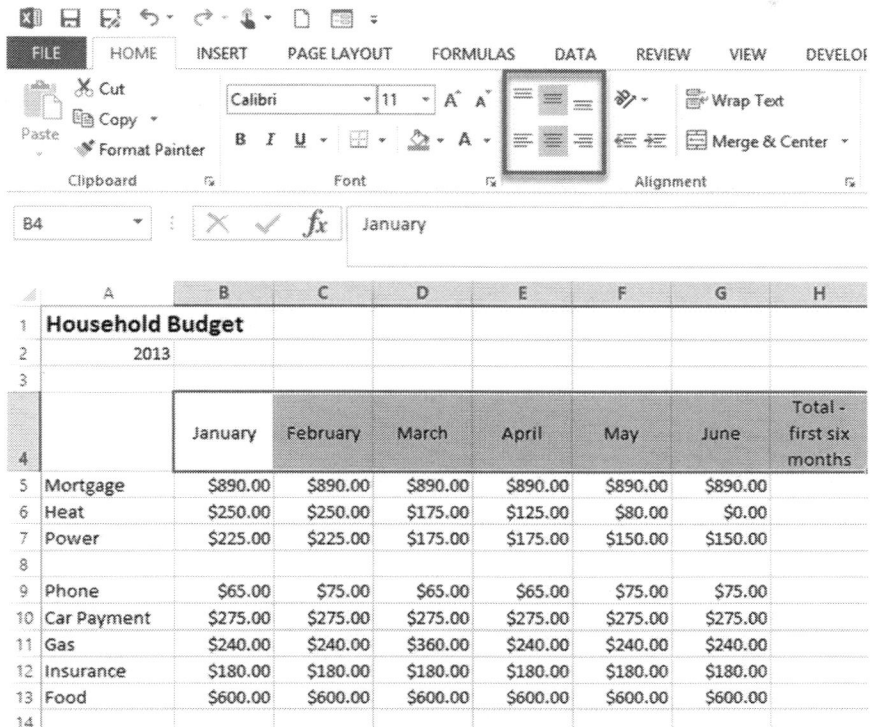

	January	February	March	April	May	June	Total - first six months
Mortgage	$890.00	$890.00	$890.00	$890.00	$890.00	$890.00	
Heat	$250.00	$250.00	$175.00	$125.00	$80.00	$0.00	
Power	$225.00	$225.00	$175.00	$175.00	$150.00	$150.00	
Phone	$65.00	$75.00	$65.00	$65.00	$75.00	$75.00	
Car Payment	$275.00	$275.00	$275.00	$275.00	$275.00	$275.00	
Gas	$240.00	$240.00	$360.00	$240.00	$240.00	$240.00	
Insurance	$180.00	$180.00	$180.00	$180.00	$180.00	$180.00	
Food	$600.00	$600.00	$600.00	$600.00	$600.00	$600.00	

To merge cells, use the following procedure.

Step 1: Highlight the cell range that you want to merge.

Step 2: Select the Merge tool from the Home tab of the Ribbon.

Step 3: Select Merge & Center.

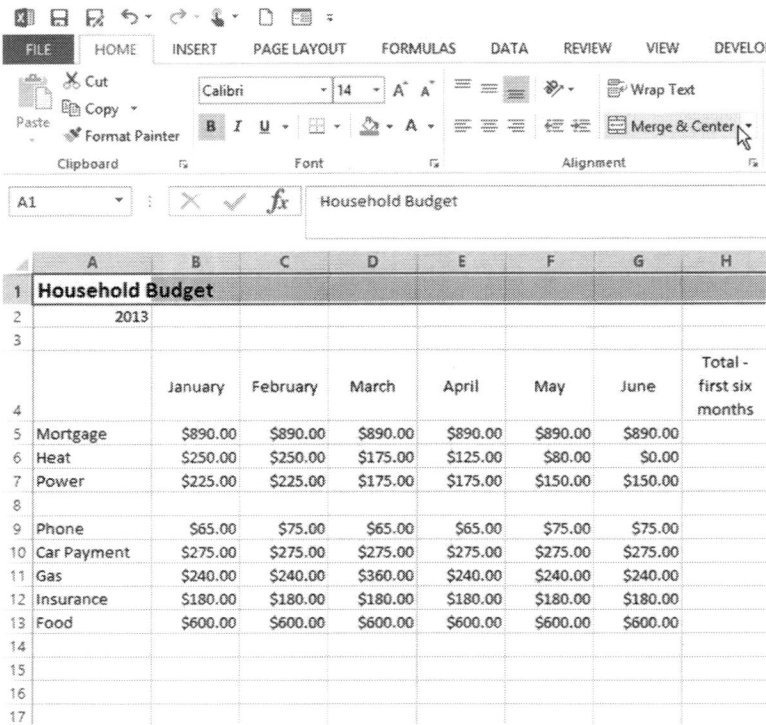

To wrap text, use the following procedure.

Step 1: Select the cell that you would like to wrap.

Step 2: Select Wrap Text from the Home tab on the Ribbon.

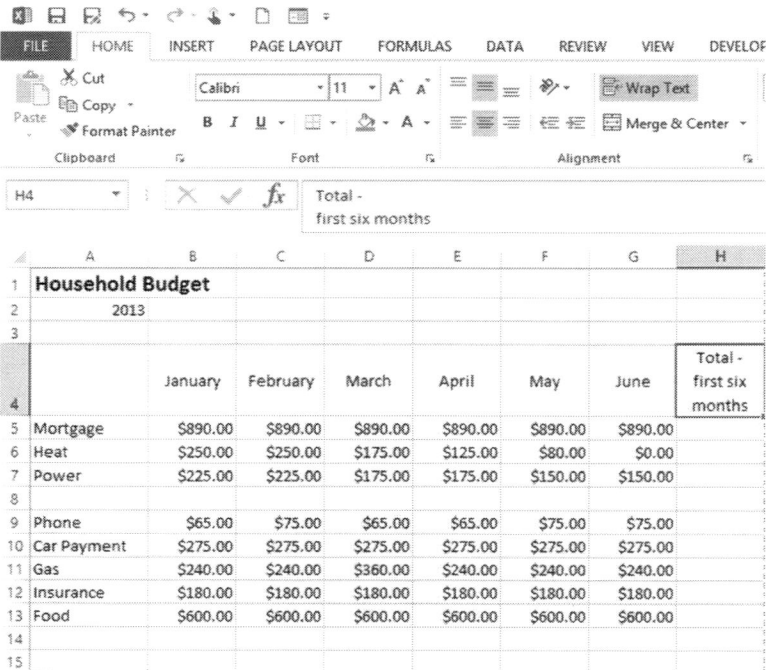

Chapter 9 – Using Styles, Themes, and Effects

In this chapter, you will learn some more advanced formatting options. First, we will discuss table styles and cell styles, including using the Format Cells dialog box to apply all types of formatting to your cells at once. This chapter explains conditional formatting. You have seen a little conditional formatting with the Quick Analysis formats. Conditional formatting is simply applying a certain type of formatting to cells that meet certain requirements. Finally, you will learn how to change the theme, colors and fonts, which can help you provide a consistent branding to all your documents, workbooks, and other office creations.

Using Table Styles and Cell Styles

To apply a table style, use the following procedure.

Step 1: Highlight the cell or cell range where you want to apply your style.

Step 2: Select the Format as Table arrow from the Home tab on the Ribbon to see the gallery of options.

Step 3: Select the style that you want to apply.

Step 4: In the Format as Table dialog box, Excel shows the cell range for your table matching your selection. If you need to change it, you can select a new range of cells. Check the My Table has Headers box if applicable. Select OK.

Format As Table ? ✕

Where is the data for your table?

=$A$4:$H$13

☑ My table has headers

OK   Cancel

The data is now formatted as a table, with filtering options in the column headers.

	Column1 ▾	January ▾	February ▾	March ▾	April ▾	May ▾	June ▾	Total – first six months ▾
1	**Household Budget**							
2	2013							
3								
4								
5	Mortgage	$890.00	$890.00	$890.00	$890.00	$890.00	$890.00	
6	Heat	$250.00	$250.00	$175.00	$125.00	$80.00	$0.00	
7	Power	$225.00	$225.00	$175.00	$175.00	$150.00	$150.00	
8								
9	Phone	$65.00	$75.00	$65.00	$65.00	$75.00	$75.00	
10	Car Payment	$275.00	$275.00	$275.00	$275.00	$275.00	$275.00	
11	Gas	$240.00	$240.00	$360.00	$240.00	$240.00	$240.00	
12	Insurance	$180.00	$180.00	$180.00	$180.00	$180.00	$180.00	
13	Food	$600.00	$600.00	$600.00	$600.00	$600.00	$600.00	
14								

To apply a cell style, use the following procedure.

Step 1: Highlight the cell or cell range where you want to apply your style.

Step 2: Select the Cell Styles tool from the Home tab of the Ribbon to see the style gallery.

Step 3: Select the style that you want to apply. You can see a preview before you select a style.

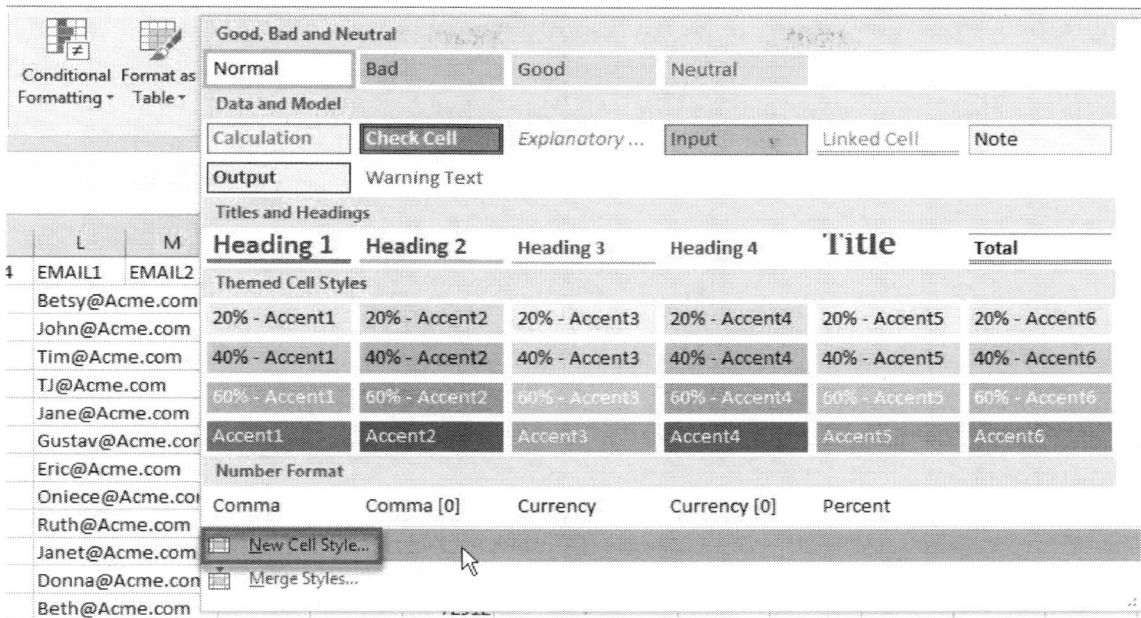

To create a new cell style, use the following procedure.

Step 1: Highlight the cell or cell range where you want to apply your style.

Step 2: Select the Cell Styles tool from the Home tab of the Ribbon to see the style gallery.

Step 3: Select New Cell Style to open the Style dialog box.

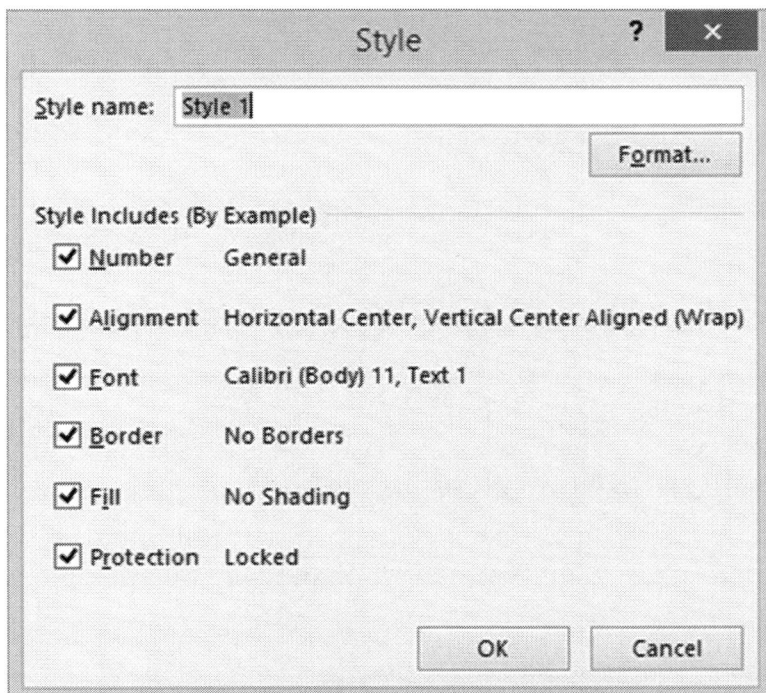

Step 4: Enter a name for the style in the Style Name field.

Step 5: Check the Style Includes boxes to indicate what formatting features the style should include.

Step 6: Select Format to open the Format Cells dialog box.

Step 7: Use the Format Cells dialog box to indicate each formatting feature for the style. Select OK when you have finished indicating all the formatting features for the style.

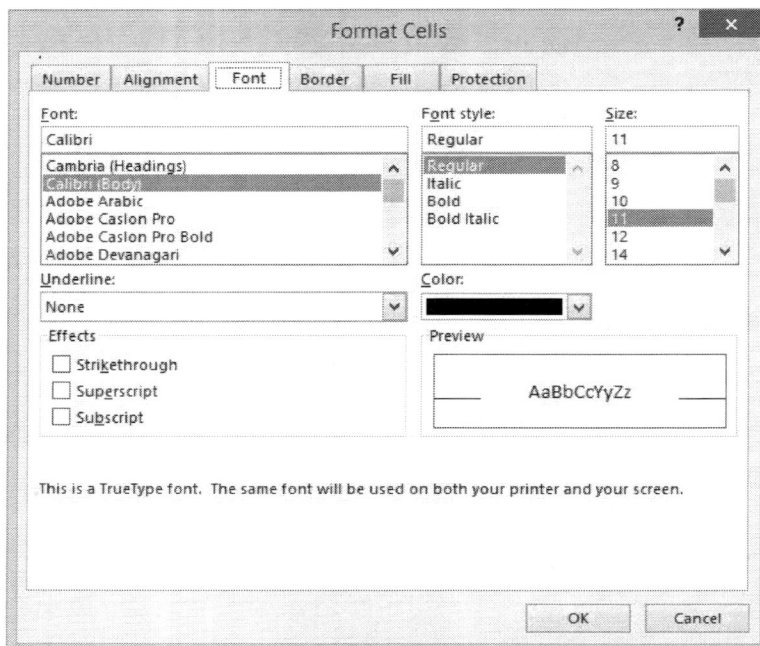

Step 8: The Number tab allows you to set number formatting for cells that contain values.

Step 9: The Alignment tab allows you to set text alignment for cells that contain text.

Step 10: The Font tab allows you to set the font for the style.

Step 11: The Border tab allows you to set customized borders for the style.

Step 12: The Fill tab allows you to set customized fill color for the style.

Step 13: The Protection tab allows you to protect the cells from changes if you use the Protection feature.

Step 14: Select OK to save your style and close the Style dialog box.

The new style appears at the top of the Cell Style gallery.

Using Conditional Formatting

To apply conditional formatting, use the following procedure. In this example, we will format all monthly totals in the budget that are over $2500.

Step 1: Highlight the cell or cell range where you want to use conditional formatting.

Step 2: Select the Conditional Formatting tool from the Home tab on the Ribbon.

Step 3: Select Highlight Cell Rules. Select Greater Than.

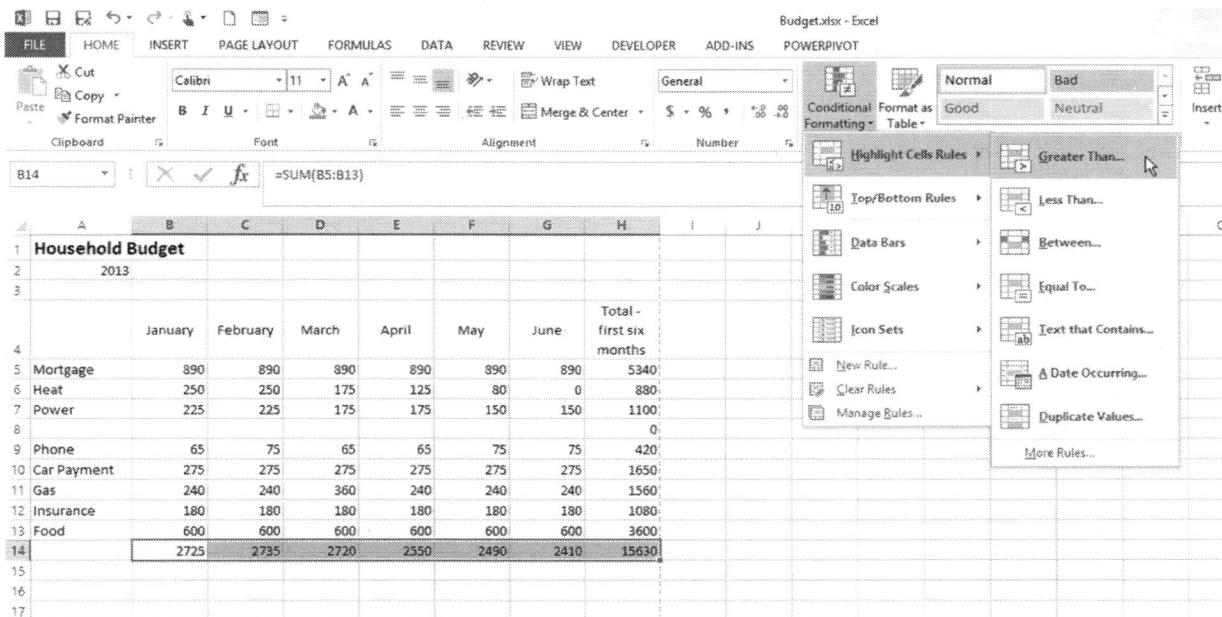

Excel displays the Greater Than dialog box to help you complete the conditional formatting rule.

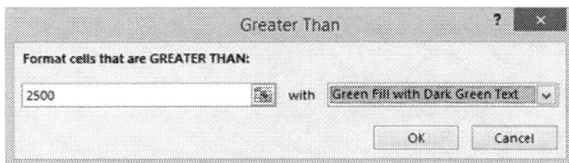

Step 4: Enter 2500 in the left field.

Step 5: Select a formatting option from the right drop down list.

Step 6: Select OK to apply the conditional formatting.

	A	B	C	D	E	F	G	H
1	**Household Budget**							
2	2013							
3								
4		January	February	March	April	May	June	Total - first six months
5	Mortgage	890	890	890	890	890	890	5340
6	Heat	250	250	175	125	80	0	880
7	Power	225	225	175	175	150	150	1100
8								0
9	Phone	65	75	65	65	75	75	420
10	Car Payment	275	275	275	275	275	275	1650
11	Gas	240	240	360	240	240	240	1560
12	Insurance	180	180	180	180	180	180	1080
13	Food	600	600	600	600	600	600	3600
14		2725	2735	2720	2550	2490	2410	15630
15								

To create a new conditional formatting rule, use the following procedure.

Step 1: Highlight the cell or cell range where you want to use conditional formatting.

Step 2: Select the Conditional Formatting tool from the Home tab on the Ribbon.

Step 3: Select New Rule.

Excel opens the New Formatting Rule dialog box.

New Formatting Rule

Select a Rule Type:

- ► Format all cells based on their values
- ► Format only cells that contain
- ► Format only top or bottom ranked values
- ► Format only values that are above or below average
- ► Format only unique or duplicate values
- ► Use a formula to determine which cells to format

Edit the Rule Description:

Format values that are:

| above ˅ | the average for the selected range |

Preview: | No Format Set | | Format... |

OK   Cancel

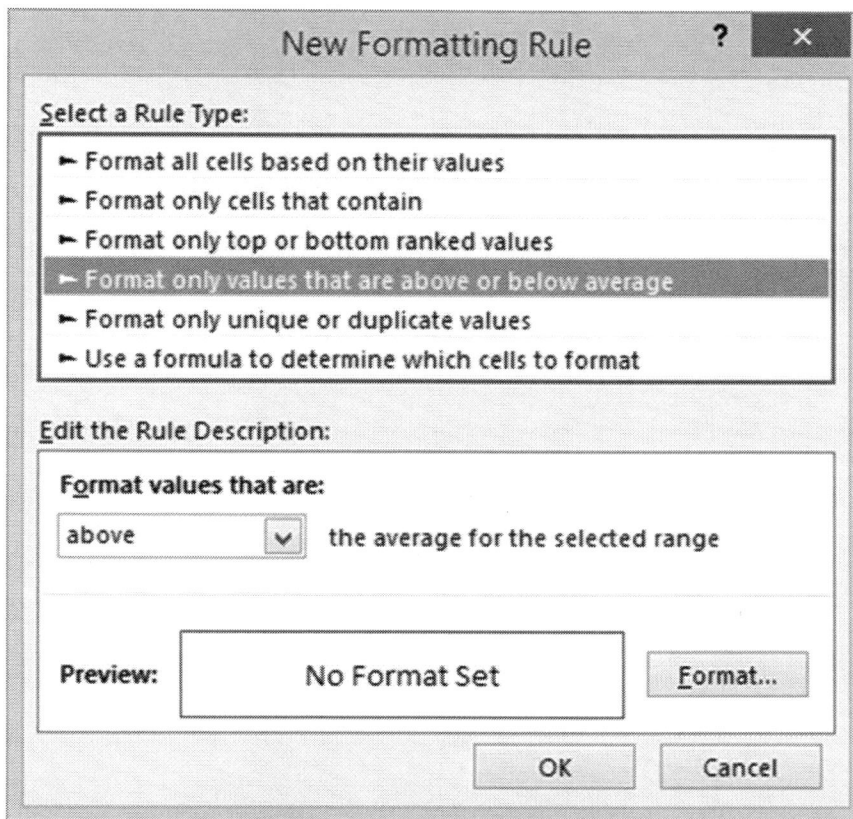

Step 4: The options in this dialog box differ, based on the Rule Type you select. Select the Rule Type and follow the prompts to indicate the conditions for when to apply the formatting.

Step 5: Select Format to open the Format Cells dialog box to create the formatting to apply when the conditions are met.

Step 6: Select OK to save your rule and close the New Formatting Rule dialog box.

Changing the Theme, Colors, and Fonts

To change the theme, use the following procedure.

Step 1: Select the Page Layout tab from the Ribbon.

Step 2: Select Themes.

Step 3: Select a new theme from the gallery. You can see a preview of each theme before you apply it.

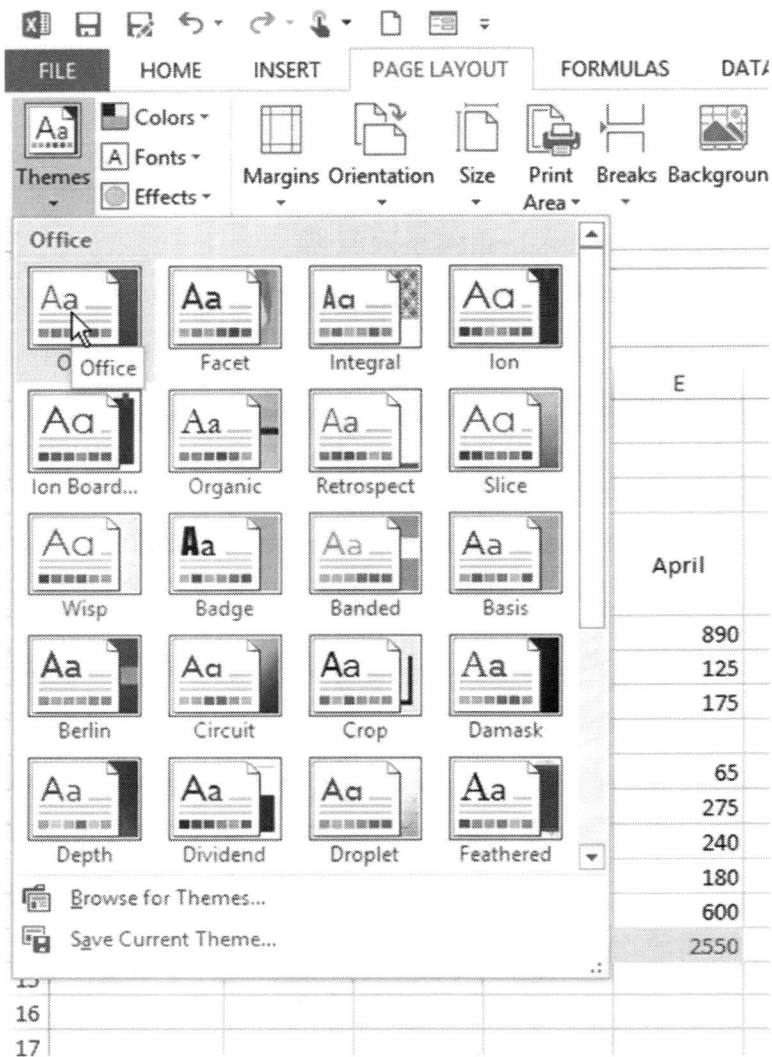

To change the colors, use the following procedure.

Step 1: Select the Page Layout tab from the Ribbon.

Step 2: Select Colors.

Step 3: Select a new color scheme from the gallery. You can see a preview of each color scheme before you apply it.

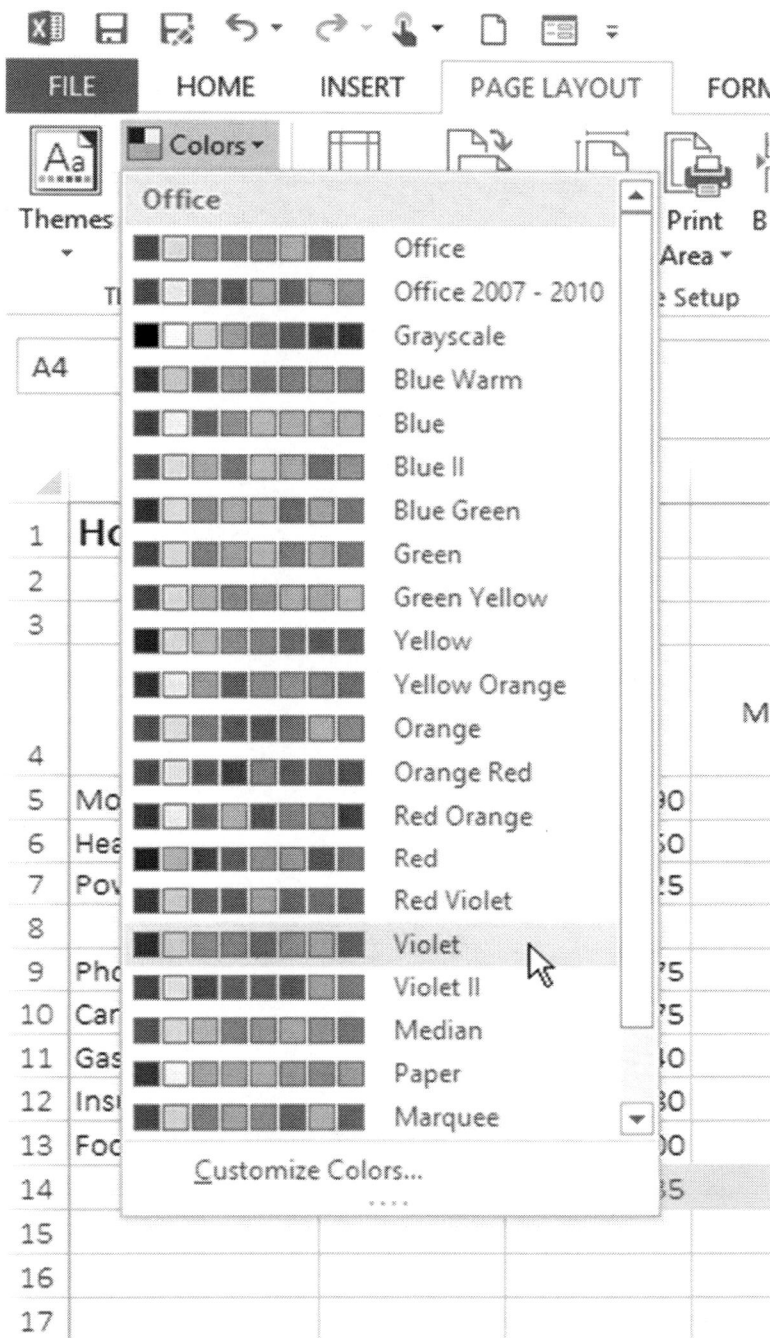

To change the fonts, use the following procedure.

Step 1: Select the Page Layout tab from the Ribbon.

Step 2: Select Fonts.

Step 3: Select a new font scheme from the gallery. You can see a preview of each font scheme before you apply it.

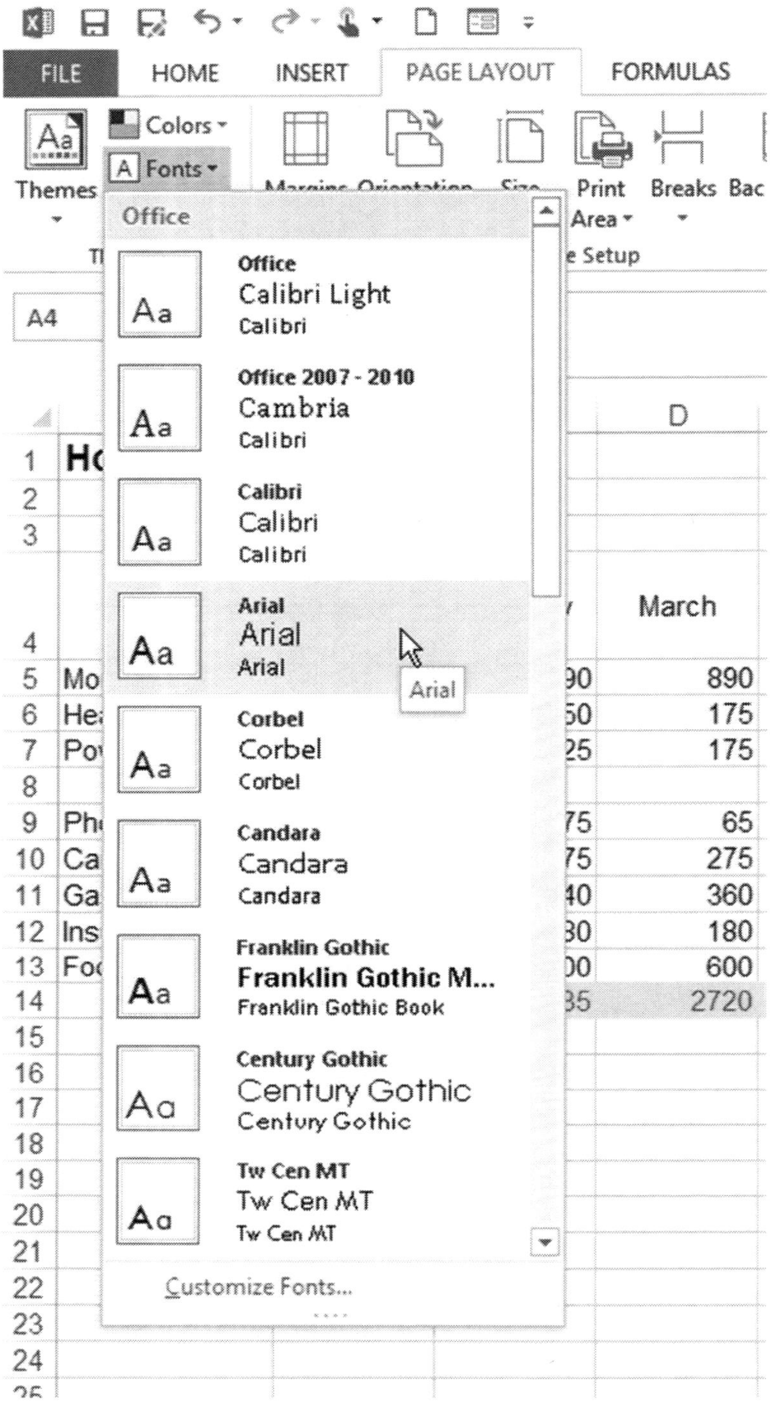
FILE
HOME
INSERT
PAGE LAYOUT
FORMULAS
Colors
Fonts
Themes
Print Area
Breaks
Office
Office
Calibri Light
Calibri
Office 2007 - 2010
Cambria
Calibri
Calibri
Calibri
Calibri
Arial
Arial
Arial
Arial
Corbel
Corbel
Corbel
Candara
Candara
Candara
Franklin Gothic
Franklin Gothic M...
Franklin Gothic Book
Century Gothic
Century Gothic
Century Gothic
Tw Cen MT
Tw Cen MT
Tw Cen MT
Customize Fonts...
A4
D
March
890
175
175
65
275
360
180
600
2720

Chapter 10 – Printing and Sharing Your Workbook

This chapter discusses printing your worksheets. First, the chapter covers the Page Layout tab for setting up the worksheet page. Next, the chapter goes into more detail on setting up your pages. The chapter discusses how to use Print Preview. Finally, the chapter explains how to print your worksheets.

An Overview of the Page Layout Tab

Explore the Page Layout tab on the Ribbon.

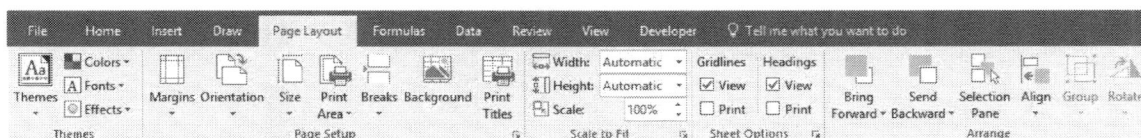

Setting up Your Page

To change the page orientation, use the following procedure.

Step 1: Select the Page Layout tab from the Ribbon.

Step 2: Select Orientation.

Step 3: Select Landscape.

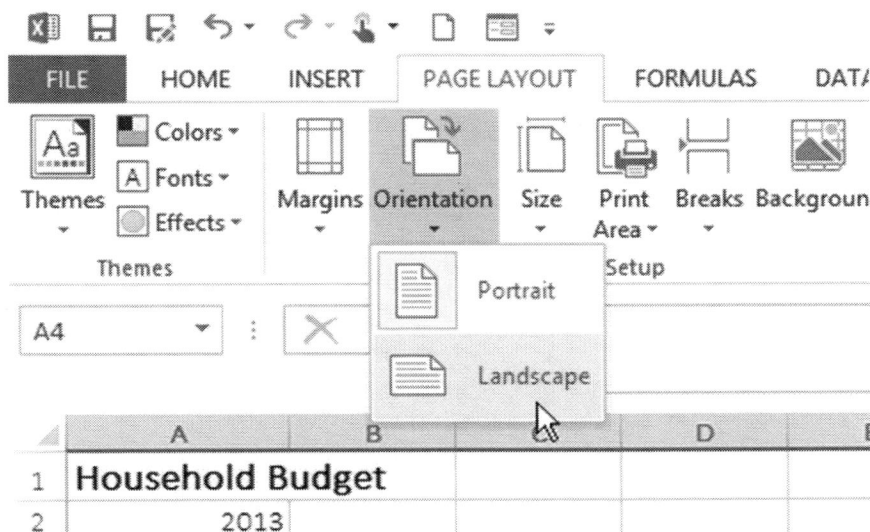

To use custom margins, use the following procedure.

Step 1: Select Custom Margins from the Margins tool on the Page Layout tab of the Ribbon.

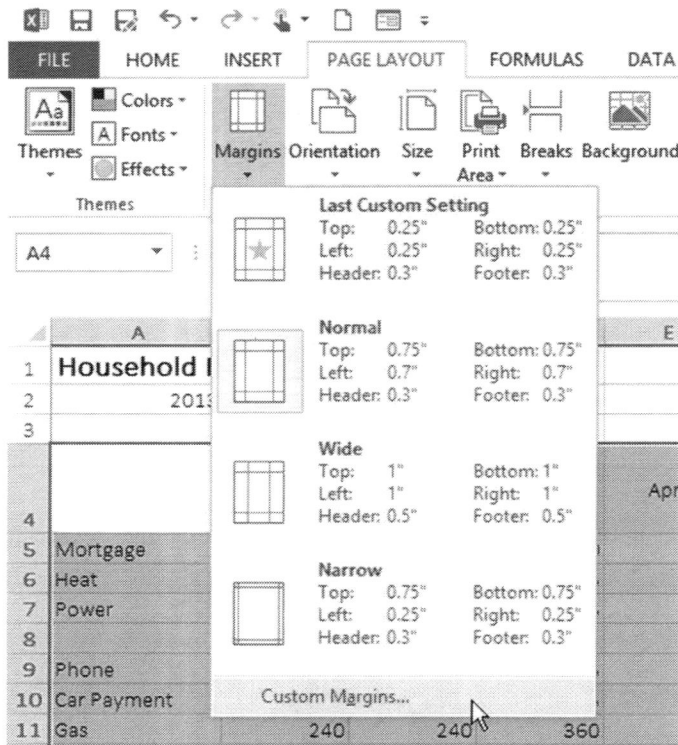

Excel displays the Page Setup dialog box.

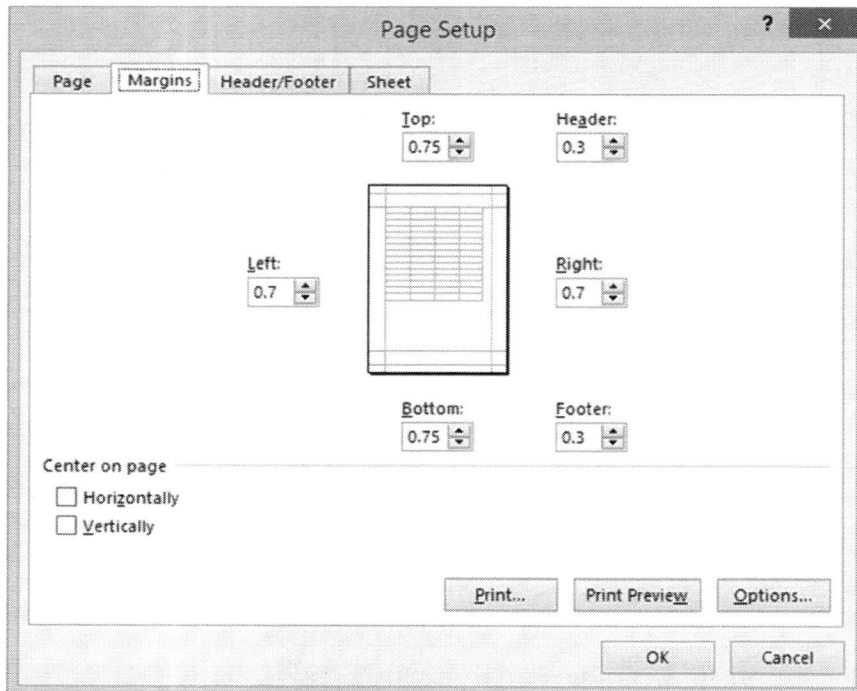

Step 2: Use the up and/or down arrows to control each of the margins (in inches). When have finished, select OK.

To insert a page break, use the following procedure.

Step 1: Place your cursor on the row or column where you want the page break to occur. Excel will insert the break above and to the left of your cursor.

Step 2: Select the Breaks tool from the Page Layout tab on the Ribbon.

Step 3: Select Insert Page Break.

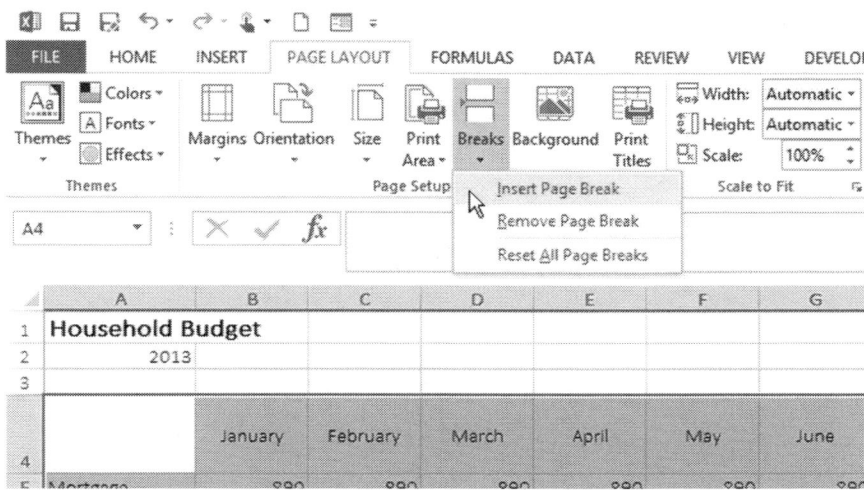

Previewing and Printing Your Workbook

Explore the Print tab in the Backstage View.

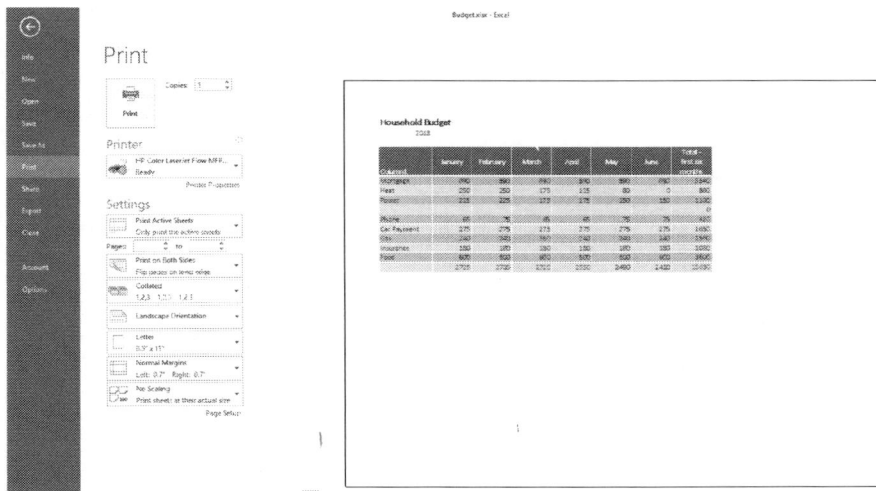

To invite people to the workbook, use the following procedure.

Step 1: Select the File tab from the Ribbon to open the Backstage view.

Step 2: Select the Share tab.

Step 3: Select Invite People.

Step 4: Enter the names or email addresses for the people that you want to invite.

Step 5: Enter a message to include with the invitation.

Step 6: If desired, check the Require User to Sign-In Before Accessing Document box to enhance the security of your workbook.

Step 7: Select Share with People.

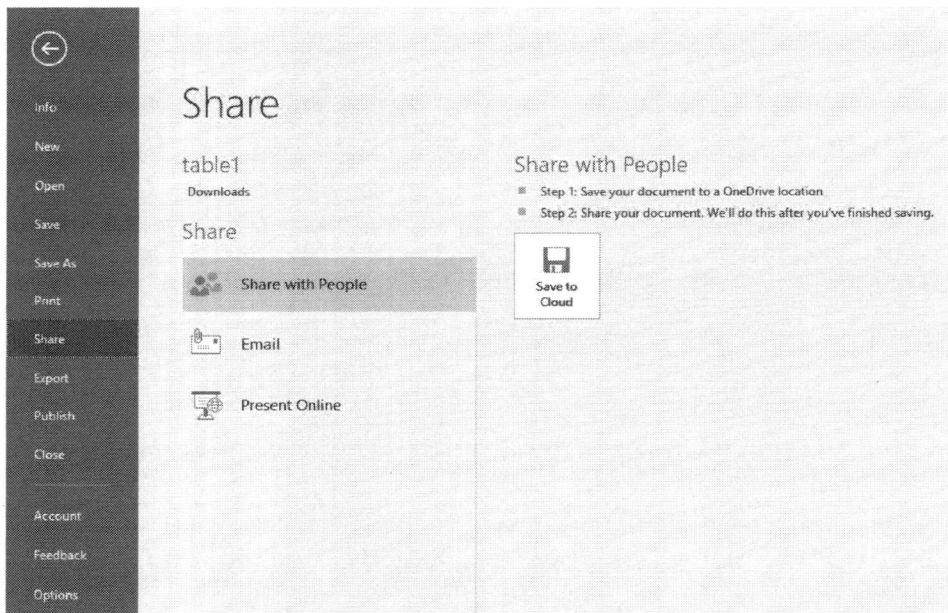

To get a link for the document, use the following procedure.

Step 1: Select the File tab from the Ribbon to open the Backstage view.

Step 2: Select the Share tab.

Step 3: Select Get a Link.

Step 4: Select the Create Link button next to View Link or Edit Link (or both), depending on what type of editing rights you want to provide. You can copy the link and paste it to another location, such as an email or a blog page.

Step 5: If you want to remove the sharing rights, select Disable Link.

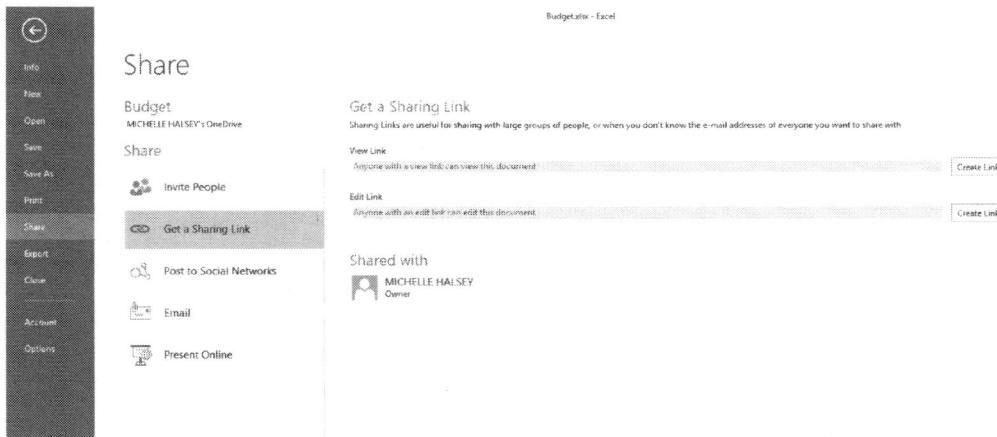

E-Mailing Your Workbook

To email an attachment or send a link, use the following procedure.

Step 1: Select the File tab on the Ribbon.

Step 2: Select the Share tab in the Backstage View.

Step 3: Select Email.

Step 4: Select Send as Attachment or Send a Link.

Step 5: Outlook opens with an email started.

If you select Send as Attachment, the name of the document is used as the subject and the document is already attached to the email. Enter the email addresses and any personal message you want to include.

If you select Send a Link, the name of the document is used as the subject and the link is included in the body message of the email. Enter the email addresses and any personal message that you want to include.

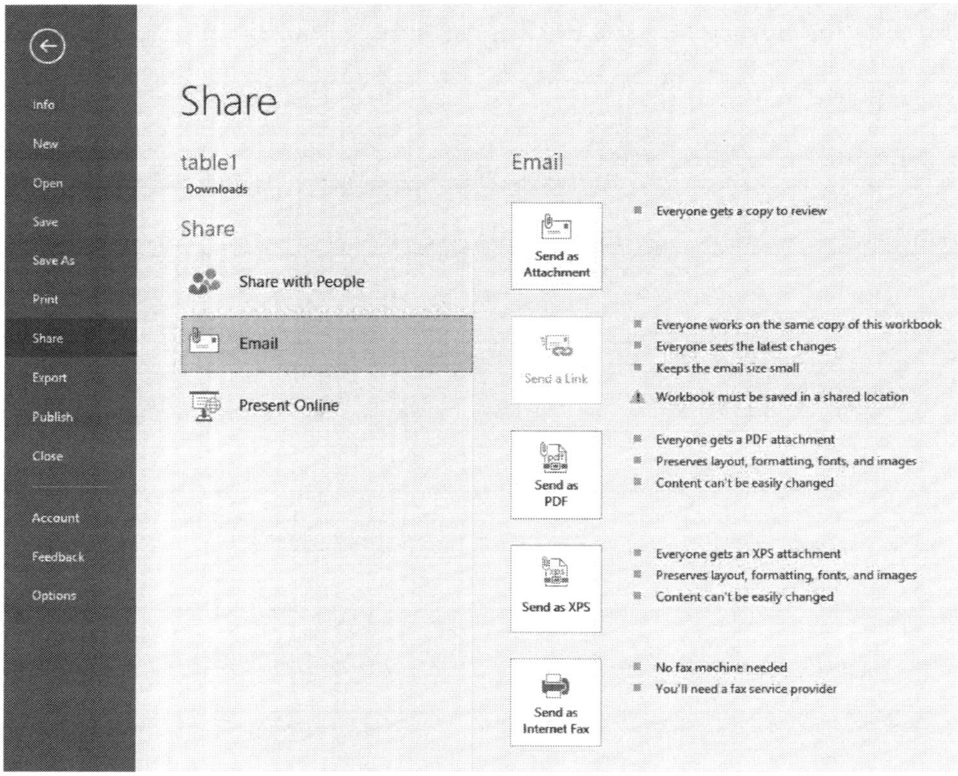
Info
New
Open
Save
Save As
Print
Share
Export
Publish
Close
Account
Feedback
Options
Share
table1
Downloads
Share
Share with People
Email
Present Online
Email
Send as Attachment
Everyone gets a copy to review
Send a Link
Everyone works on the same copy of this workbook
Everyone sees the latest changes
Keeps the email size small
Workbook must be saved in a shared location
Send as PDF
Everyone gets a PDF attachment
Preserves layout, formatting, fonts, and images
Content can't be easily changed
Send as XPS
Everyone gets an XPS attachment
Preserves layout, formatting, fonts, and images
Content can't be easily changed
Send as Internet Fax
No fax machine needed
You'll need a fax service provider

SmartArt, pictures, text boxes, and shapes are different ways to enhance your spreadsheet, especially when sharing the information with others. In this chapter, we will look at how to add these objects to your spreadsheets. We will also look at how to edit a SmartArt diagram. Finally, you will learn about the contextual Tools tabs that appear in Excel 2016 when you are working with different types of objects.

Inserting SmartArt

To insert SmartArt, use the following procedure.

Step 1: Select the Insert tab from the Ribbon.

Step 2: Select SmartArt.

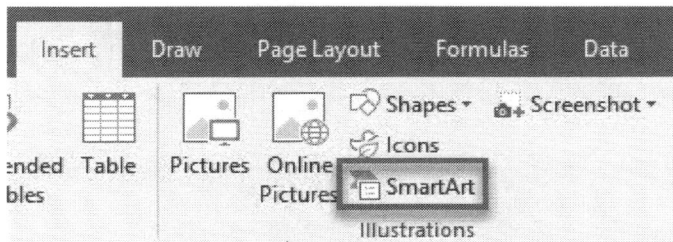

Step 3: In the Choose a SmartArt Graphic dialog box, select the category on the left. Then you select the item in the middle. The right shows a preview of the item. Select OK to insert the content.

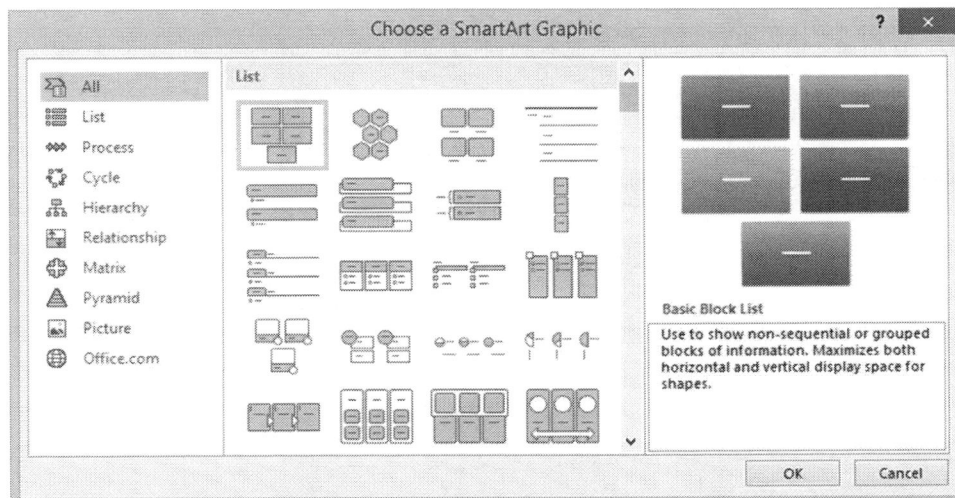

Excel inserts the selected SmartArt graphic in the middle of the spreadsheet.

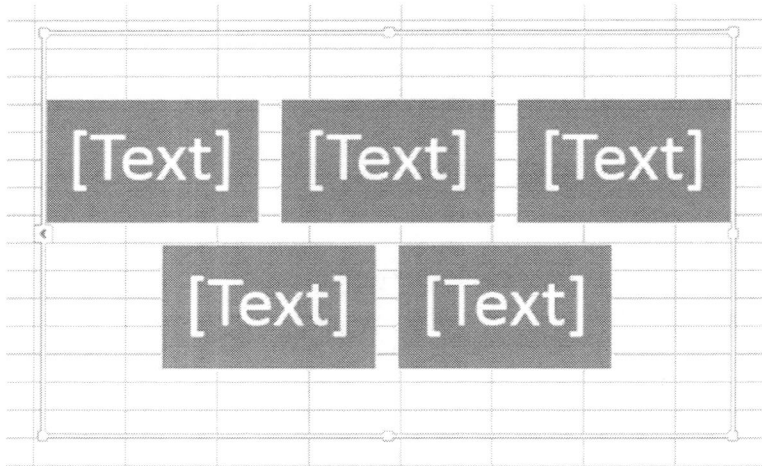

You can simply click on one of the boxes and type in your text, if desired. Notice that the font sizes adjust, depending on how much text you enter.

To add text to a SmartArt graphic using the Text pane, use the following procedure.

Step 1: To the left of the SmartArt graphic you inserted, there is a small rectangle with an arrow. Click this arrow to open the Text Pane.

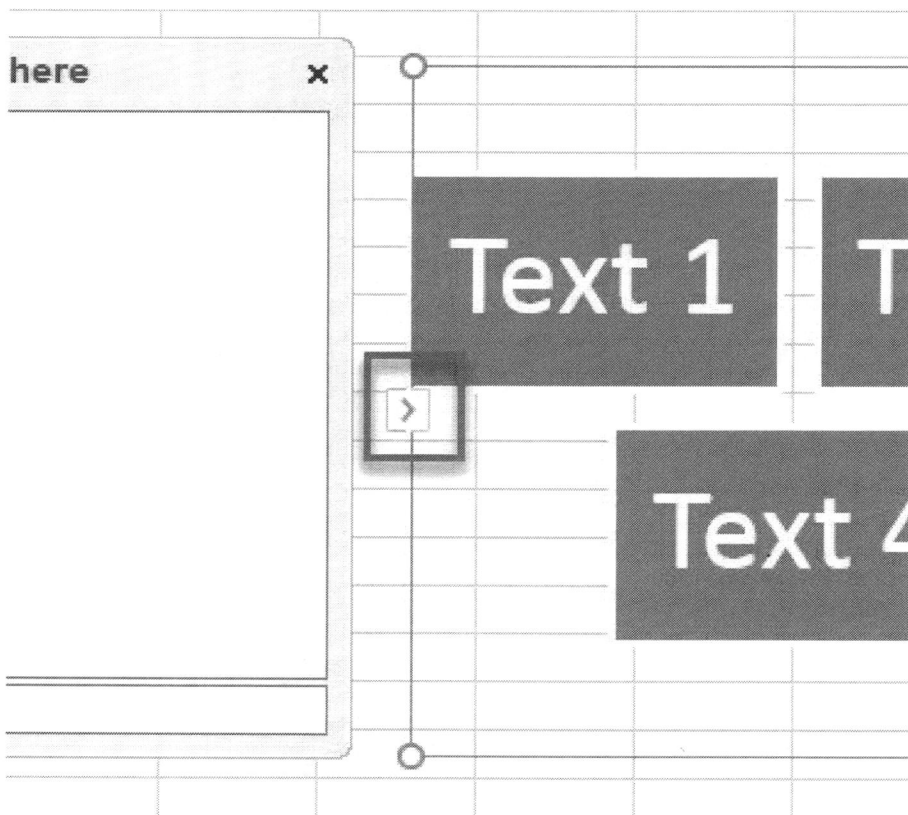

Excel opens the Text Pane.

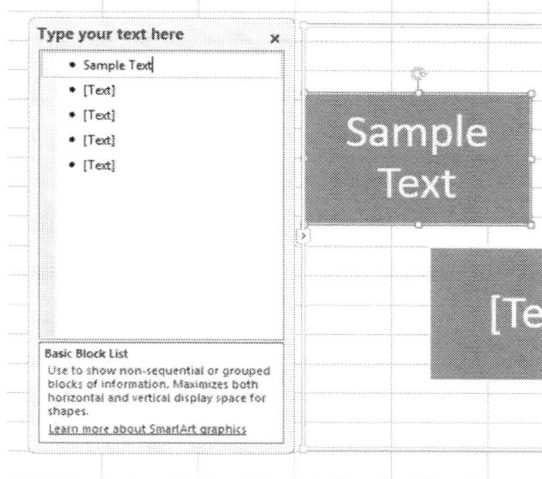

Step 2: Click on the first line and begin typing. Each line represents a new item in the graphic.

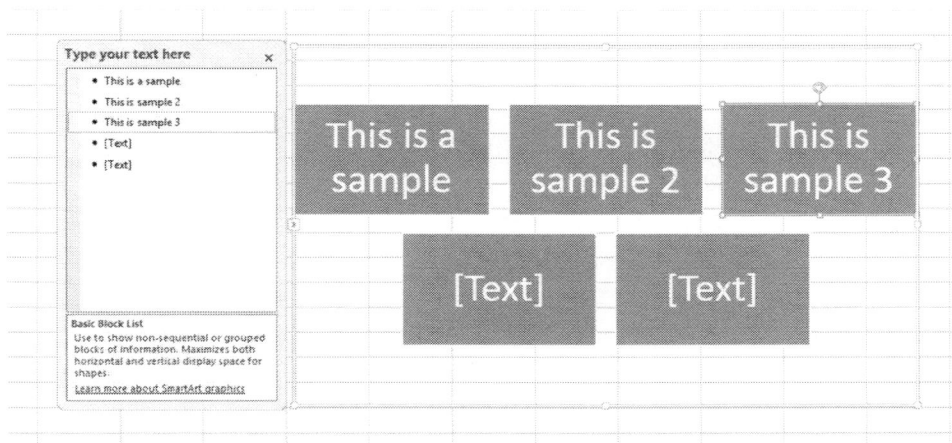

The SmartArt text adjusts to fit the graphic. The more text you enter in each graphic element, the smaller the text will become.

Step 3: When you have finished, click anywhere on the spreadsheet, and the Text Pane will close automatically. Or you can click the X in the top right corner.

Editing the Diagram

To resize a SmartArt graphic, use the following procedure.

Step 1: Select the SmartArt graphic to select it. Notice the border around the graphic.

Step 2: Select one of the corners and drag the picture. Notice the cursor changes to a diagonal line with arrows at both ends. You can make it smaller or bigger, depending on which direction you drag.

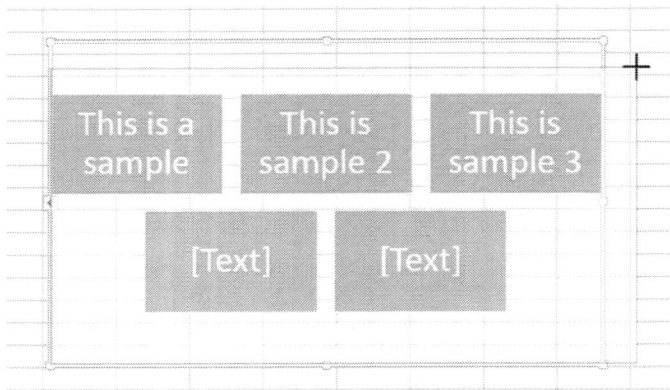

Step 3: Release the mouse when the graphic is the desired size. Notice that Excel may rearrange the graphic elements for the best look and fit.

To move the diagram, use the following procedure.

Step 1: Select the diagram border.

The cursor changes to a cross with four arrows.

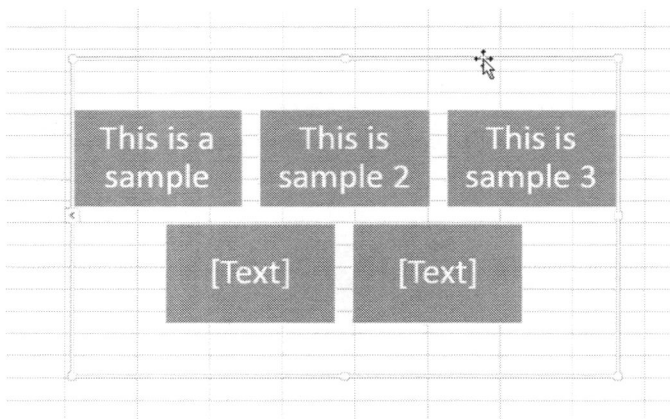

Step 2: Drag the mouse to move the diagram. Release the mouse when the diagram is in the desired location.

Note that you can move the individual parts of the SmartArt diagram using the same procedure. Just click on the individual object you want to move. Practice this for the next segment of the lesson.

Resetting the diagram allows you to quickly return the graphic to the original alignment and spacing between elements, use the following procedure to reset a diagram.

Step 1: Right-click on the diagram.

Step 2: Select Reset Graphic from the context menu.

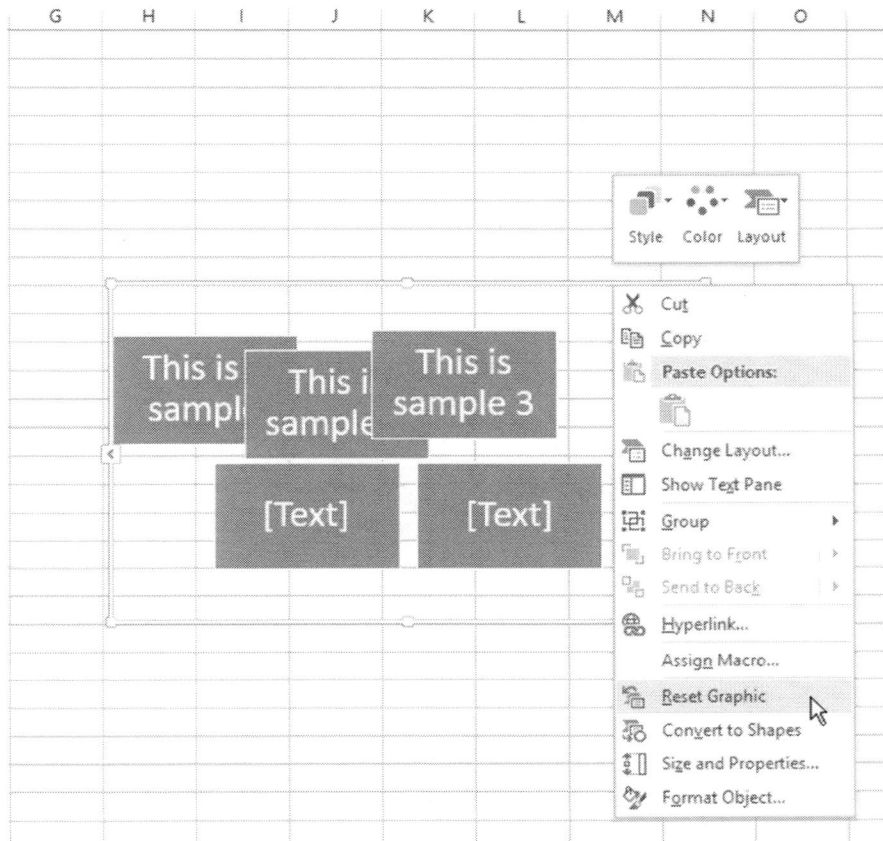

Adding Pictures

To insert a picture from a file, use the following procedure.

Step 1: Select the Insert tab from the Ribbon.

Step 2: Select Picture.

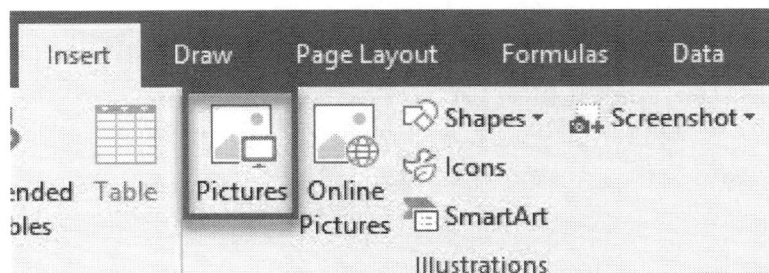

Step 3: Navigate to the location of the file and highlight the file you want to insert.

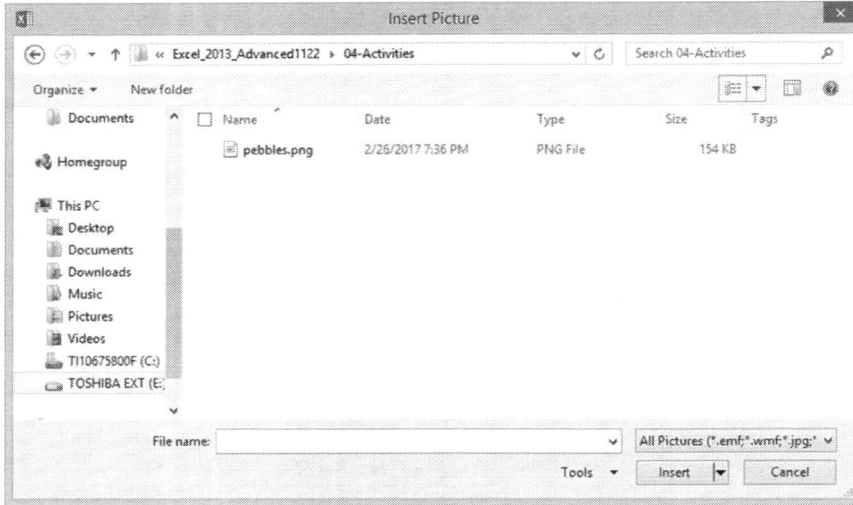

Step 4: Select Insert.

Excel inserts the picture.

To insert an online picture, use the following procedure.

Step 1: Select the Insert tab from the Ribbon.

Step 2: Select Online Pictures.

Step 3: Enter a search term. Press Enter to begin searching.

Step 4: Excel displays the matching images. To insert one, double-click it or highlight it and select Insert.

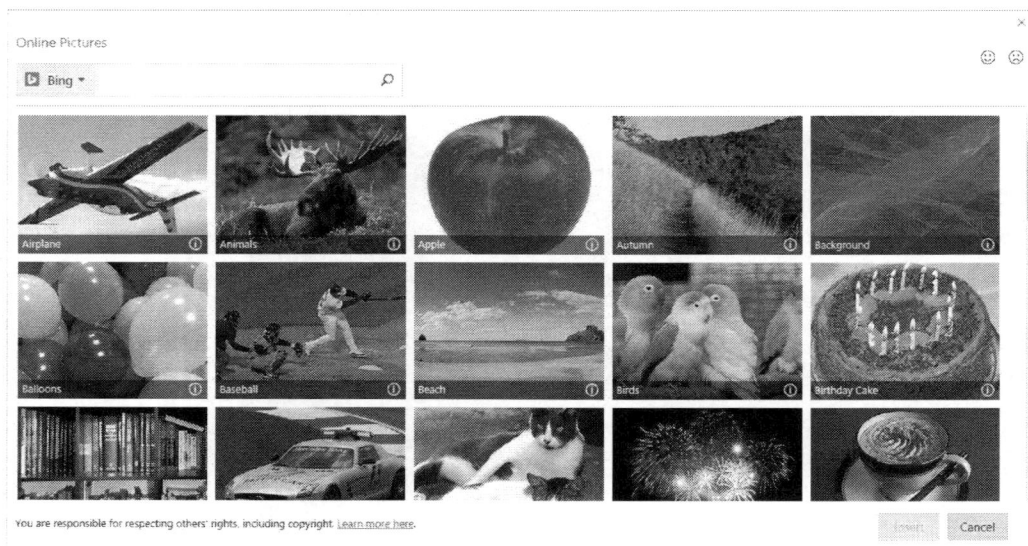

To insert a full-size screenshot, use the following procedure.

Step 1: Select the Insert tab from the Ribbon.

Step 2: Select Screenshot.

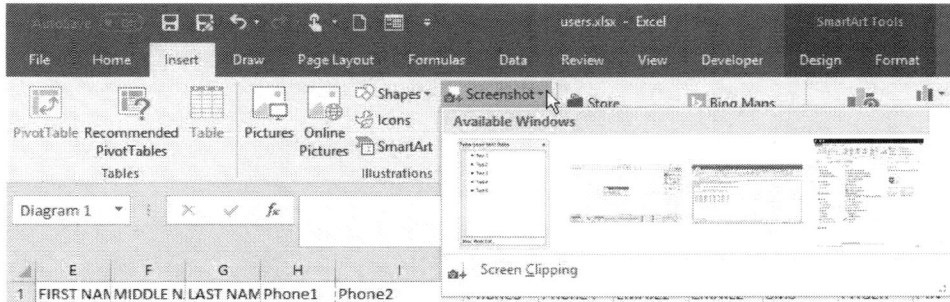

Step 3: The Screenshot gallery includes a thumbnail image of other windows you have open. Select the image that you want to insert.

Excel inserts the image.

To insert a screen clipping, use the following procedure.

Step 1: Make sure that the area of the screen you want is ready to capture. Excel will automatically return to the previous window for a screen clipping.

Step 2: Select the Insert tab from the Ribbon.

Step 3: Select Screenshot.

Step 4: Select Screen Clipping.

Step 5: Drag the mouse to capture the area of the screen that you want to insert in your presentation. The screen is slightly greyed out, except for the area you are capturing.

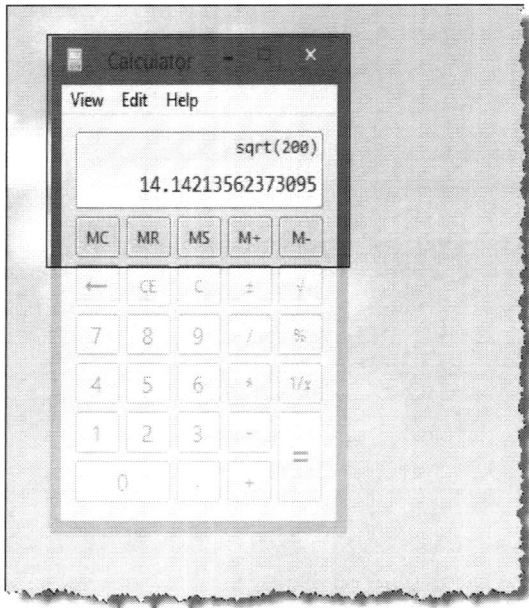

Step 6: When you release your mouse, Excel inserts the screen clipping into the workbook at the current cursor position.

Adding Text Boxes

To insert a text box, use the following procedure.

Step 1: Select the Insert tab from the Ribbon.

Step 2: Select Text Box. Select Horizontal Text Box or Vertical Text Box.

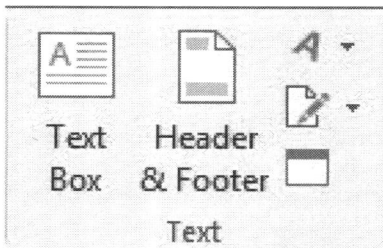

Step 3: Click on the worksheet and drag the mouse to draw the text box.

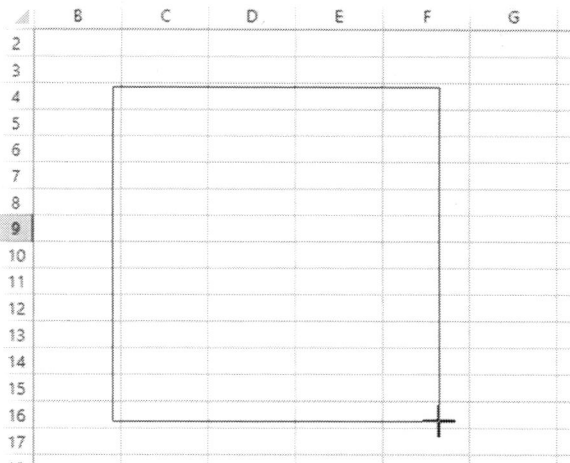

Step 4: When you release the mouse, Excel inserts the text box.

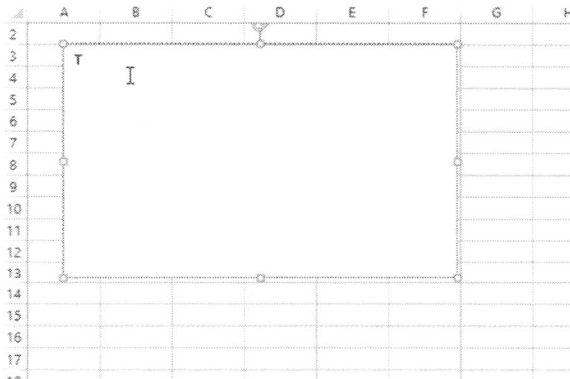

Step 5: Begin typing to enter text into the text box.

Drawing Shapes

To draw a shape, use the following procedure.

Step 1: Select the Insert tab from the Ribbon.

Step 2: Select Shapes.

Excel displays the Shapes gallery.

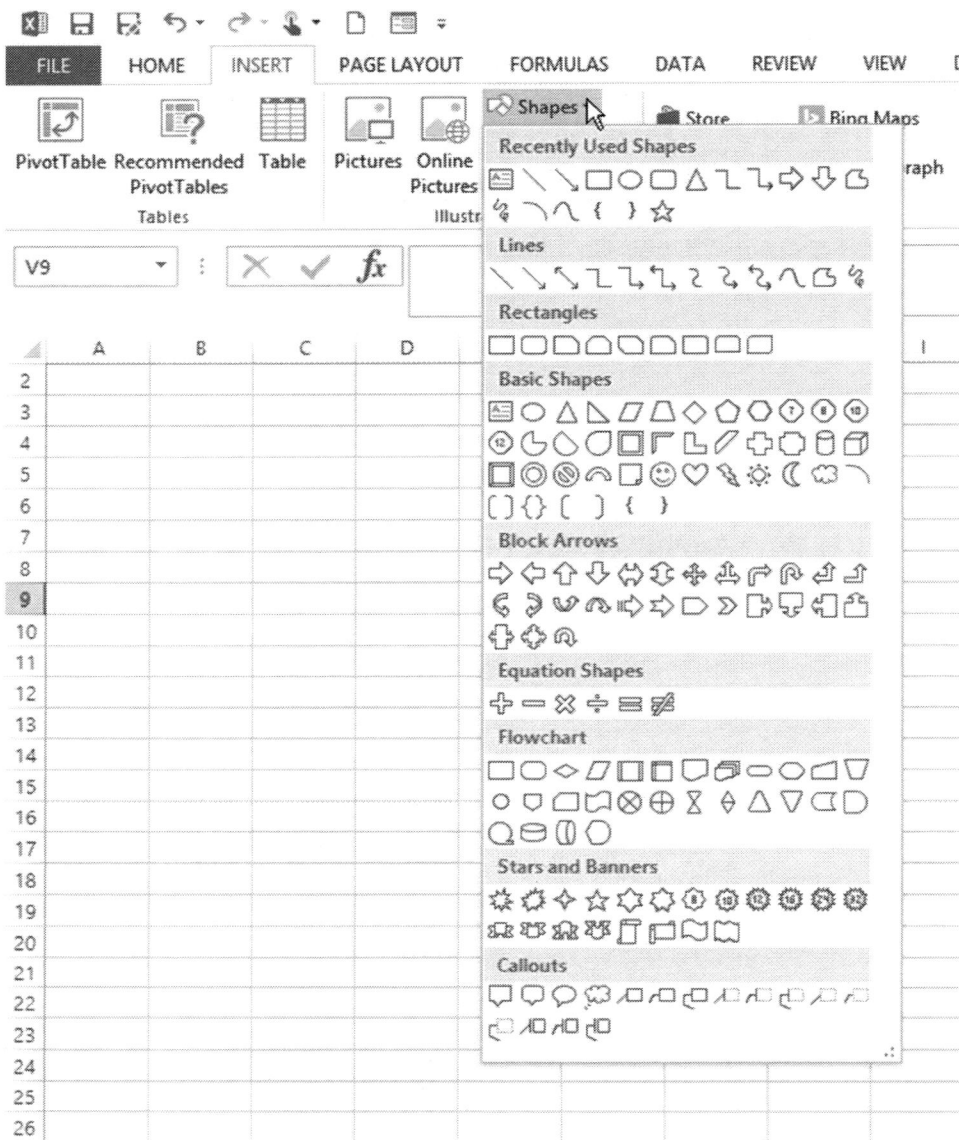

Step 3: Select a shape tool.

Step 4: Drag the mouse in the desired location to create the selected shape. The cursor is a cross while you are drawing.

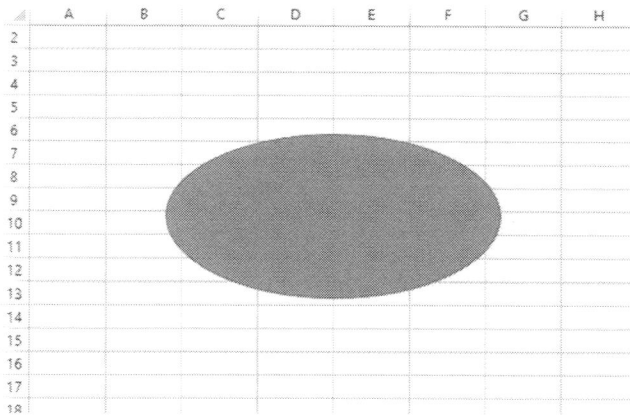

Step 5: Release the mouse to complete the shape.

About the Contextual Tabs

The Tools tabs for working with SmartArt.

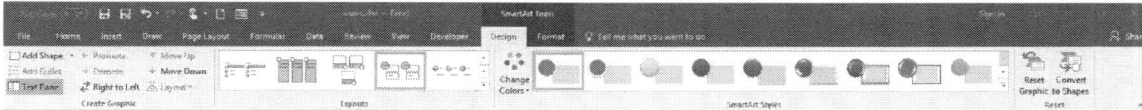

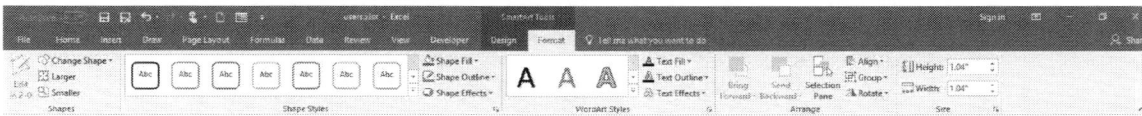

000The Tools tab for working with pictures.

The Tools tab for working with a Text box or shape.

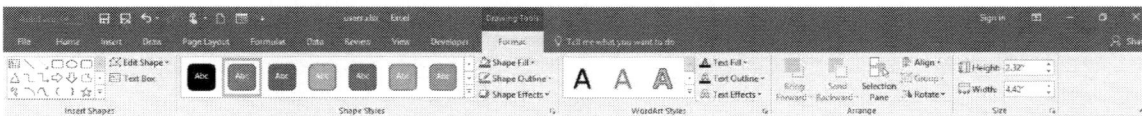

Chapter 12 – Auditing

This chapter introduces concepts that will help you troubleshoot formulas as we progress through the course. Precedent cells are cells whose contents are used in the active cell. Dependent cells are used in other cells contents or formulas. This chapter explains how to show these relationships. It also explains how to display the formulas, instead of the results, in a worksheet. You will also learn how to work with comments in this chapter.

Tracing Precedent Cells

To trace precedents, use the following procedure.

Step 1: Select the cell that contains the formula you want to trace. Cell D18 is used in this example.

Step 2: Select the Formulas tab from the Ribbon.

Step 3: Select Trace Precedents.

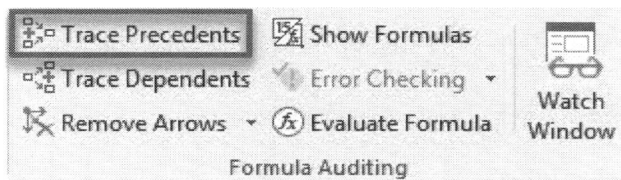

Step 4: Excel adds a tracer arrow from each cell that provides data to the active cell.

Line Item	February	March
PROFIT AND LOSS		
Revenue		
Budget	$75,000	$85,000
Actual	$70,000	$88,000
Budget variance (Actual – Budget)	($5,000)	$3,000
Prior year	$60,000	$70,000
Prior year variance (Actual – Prior year)	$10,000	$18,000
Cost of Goods Sold		
Budget	$55,000	$65,000

To remove the tracers, select Remove Arrows.

Tracing the Dependents of a Cell

To trace dependents, use the following procedure.

Step 1: Select the cell that you want to trace. Cell D18 is used in this example.

Step 2: Select the Formulas tab from the Ribbon.

Step 3: Select Trace Dependents.

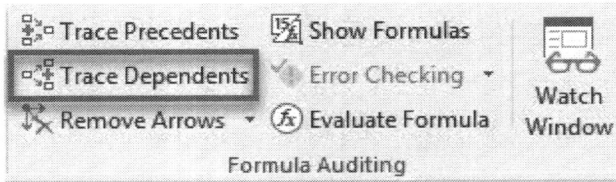

Step 4: Excel adds a tracer arrow to each cell that uses the active cell's data.

Step 5: Click the Trace Dependents tool again to see further relationships that are influenced by the active cell's contents.

Line Item	February	March	Q1	April
PROFIT AND LOSS				
Revenue				
Budget	$75,000	$85,000	$225,000	$75,00
Actual	$70,000	$88,000	$218,000	$95,00
Budget variance (Actual – Budget)	($5,000)	$3,000	($7,000)	$20,00
Prior year	$60,000	$70,000	$185,000	$75,00
Prior year variance (Actual – Prior year)	$10,000	$18,000	$33,000	$20,00
Cost of Goods Sold				

Displaying Formulas Within the Sheet

To display formulas within the sheet, use the following procedure.

Step 1: Select the Formulas tab from the Ribbon.

Step 2: Select Show Formulas.

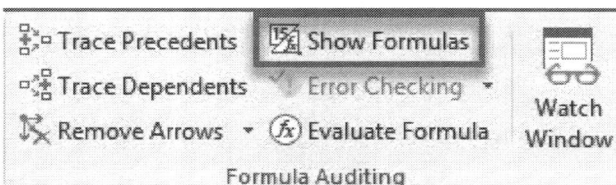

Excel expands the cells as necessary and displays all the worksheet's formulas in their cells.

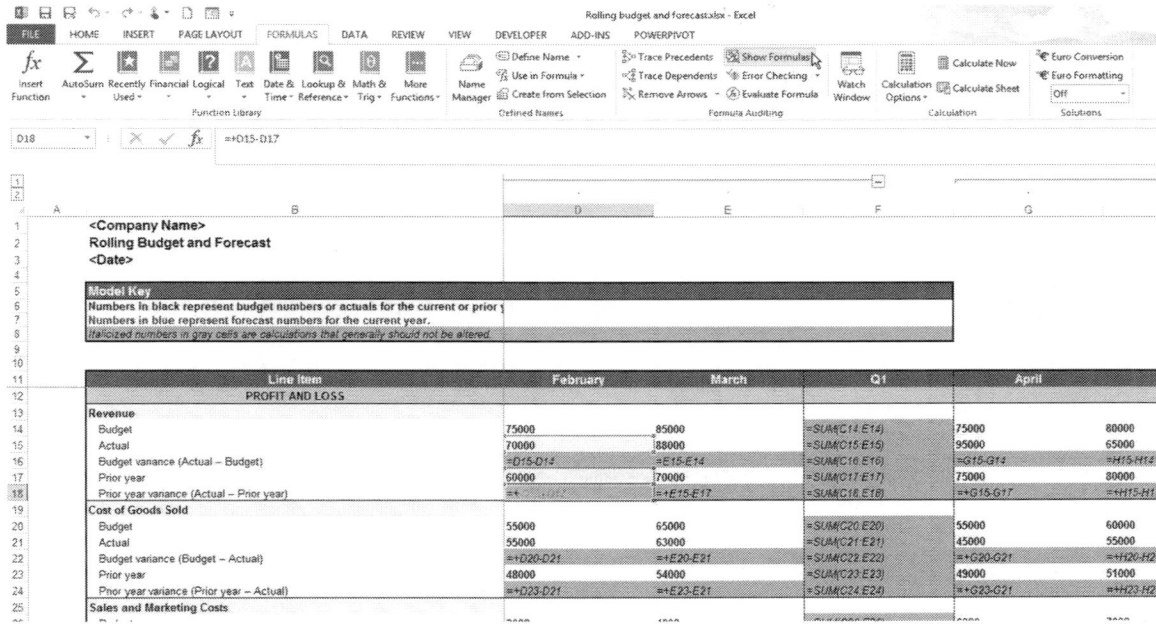

Adding, Displaying, Editing, and Removing Comments

To add a comment, use the following procedure.

Step 1: Select the cell where you want to add a comment.

Step 2: Select the Review tab from the Ribbon.

Step 3: Select New Comment.

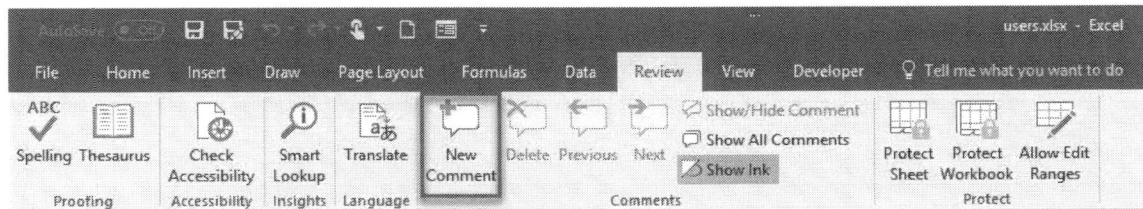

Step 4: Begin typing your comment.

	Q1	April	May
	$225,000	$75,000	$80,000
	$218,000	$95,000	$65,000
)	($7,000)		
	$185,000	Sample Comment|	
year)	$33,000		
	$170,000	$55,000	$60,000
	$165,500	$45,000	$55,000
)	$4,500	$10,000	$5,000
	$147,000	$49,000	$51,000
ctual)	($18,500)	$4,000	($4,000)

To show or hide comments, use the following procedure.

Step 1: Select the cell with the comment.

Step 2: Select Show/Hide Comment or Show All Comments.

To edit a comment, use the following procedure.

Step 1: Select the cell with the comment.

Step 2: Select Edit Comment.

Excel opens the comment for editing. You can select text to change it, delete, or add text to the comment.

Q1	April	May	Ju
$225,000	$75,000	$80,000	$90,
$218,000	$95,000	$65,000	$88,
($7,000)	Sample Comment		($2,
$185,000			$90,
$33,000			($2,
$170,000	$55,000	$60,000	$70,
$165,500	$45,000	$55,000	$63,

To remove a comment, use the following procedure.

Step 1: Select the cell with the comment.

Step 2: Select Delete from the Review tab on the Ribbon.

Chapter 13 – Creating Charts

Charts provide a visual way of relating information. We will start with a new feature in 2016: Recommended Charts. Excel provides a customized set of charts based on data you select. This chapter will also explain how to insert a chart of your choosing. You will learn about the chart tools tab and gain an overview of the parts of a chart. Finally, you will learn how to resize and move a chart.

Using Recommended Charts

To insert a recommended chart, use the following procedure.

Step 1: Select the data that you want to use in your chart.

Step 2: Select the Insert tab from the Ribbon.

Step 3: Select Recommended Charts.

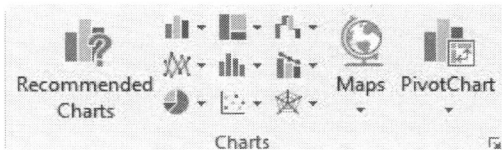

Step 4: In the Insert Chart dialog box, the Recommended Charts tab shows several charts that Excel recommended for the type of data you have selected. As you select each option on the left side of the dialog box, the right side shows a preview.

Step 5: When you find a chart that you want to use, select it in the list, and select OK.

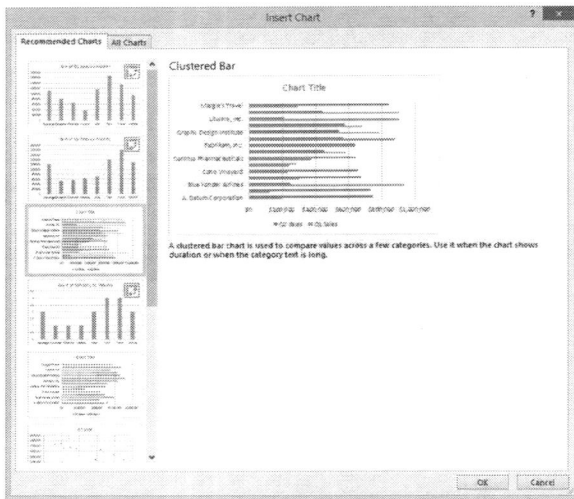

Inserting a Chart

The types of charts in Excel 2016 are:

Column charts – Column charts allow you to visually compare values across a few categories.

Line Chart – Line charts show trends over time (such as years, months, or days) or categories.

Pie charts – Pie charts show your data in proportions of a whole. The total of your numbers should be 100%. Doughnut charts are included. Donut charts are good when there are multiple series that relate to a larger sum.

Bar charts – Bar charts allow you to visually compare values across a few categories when the chart shows duration or the category text is long.

Area charts – Area charts show trends over time or categories. Use it to highlight the magnitude of change over time.

Scatter charts – Scatter or bubble charts show the relationship between sets of values.

Stock, Surface or Radar charts – This category includes many chart types to help you show the trend of a stock's performance over time, show trends in values in a curve or with color, or show values relative to a center point.

Combo chart – Combo charts highlight different types of information when your values vary widely or you have mixed types of data.

To insert a chart, use the following procedure.

Step 1: Select the cells, including the labels to include in the chart.

Step 2: Select the Insert tab from the Ribbon.

Step 3: Select the type of chart you would like to use.

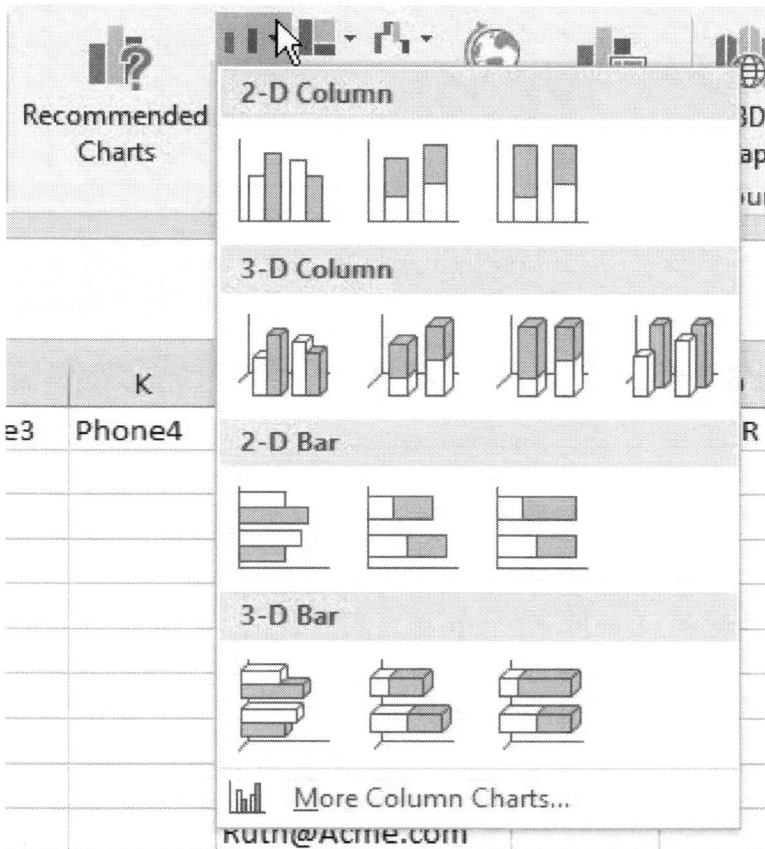

Excel displays the chart.

The Tools tabs for working with charts.

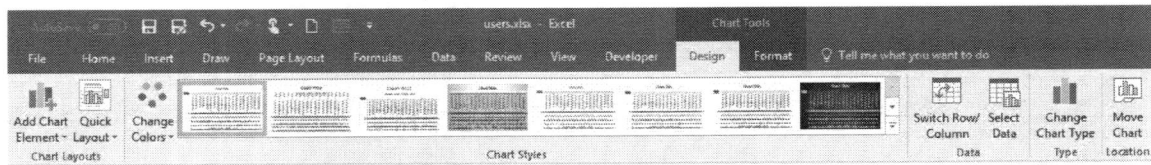

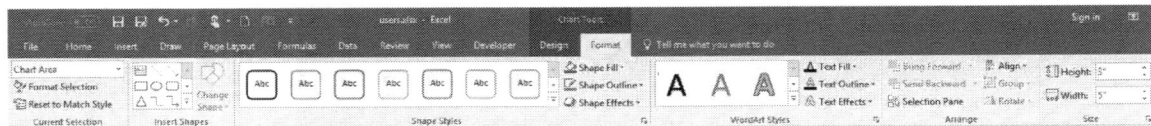

Understanding Chart Elements

The parts of a standard chart are.

- The Chart area includes all other parts of the chart that appear inside the chart window.

- A data point represents a single value in the worksheet. Depending on the type of chart, this may be a bar, a pie slice, or another shape or pattern.

- A group of related data points make up the data series. Charts usually have more than one data series, except pie charts, which only represents one data series.

- An axis is a reference line for plotting data. A two-dimensional chart has an X-axis and a y-axis. For many charts, the label is on the X-axis and the values are on the y-axis. Three dimensional charts also have a Z-axis. A pie chart does not have an axis of any type.

- A tick mark intersects an axis as a small line. It may have a label and can indicate a category, scale, or chart data series.

- The Plot area includes all axes and data point markers.

- Gridlines can make it easier to view data values by extending tick marks across the whole plot area.

- You can add text to include a label or title. The text can be attached to the chart or axis, which cannot be moved independently of the chart. Unattached text is a text box simply shown with the chart.
- The legend defines the patterns, colors, or symbols used in the data markers.

Resizing and Moving the Chart

To resize a chart, use the following procedure.

Step 1: Click on the chart to select it. Notice the border around the chart.

Step 2: Select one of the corners and drag the chart. Notice the cursor changes to a diagonal line with arrows at both ends. You can make it smaller or bigger, depending on which direction you drag.

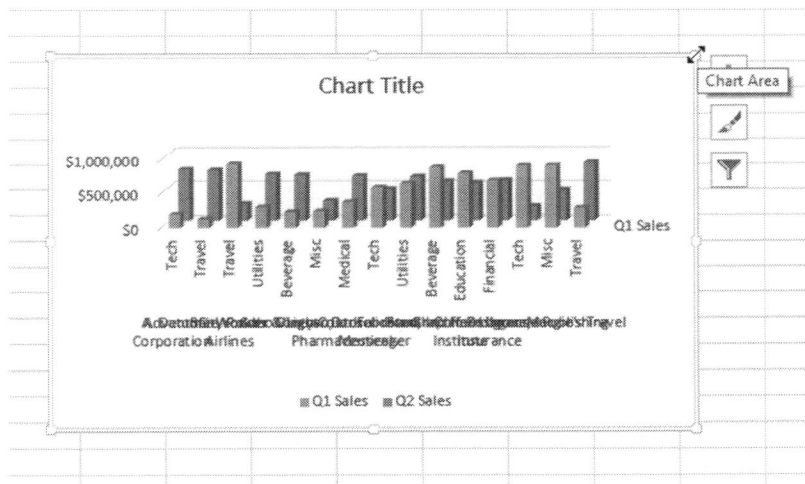

Step 3: Release the mouse when the chart is the desired size.

To move the chart to a new worksheet in the workbook, use the following procedure.

Step 1: Select the chart.

Step 2: Select the Chart Tools Design tab.

Step 3: Select the Move Chart tool.

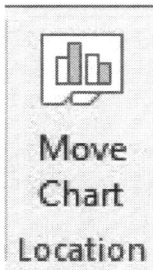

Move
Chart
Location

Excel displays the Move Chart dialog box.

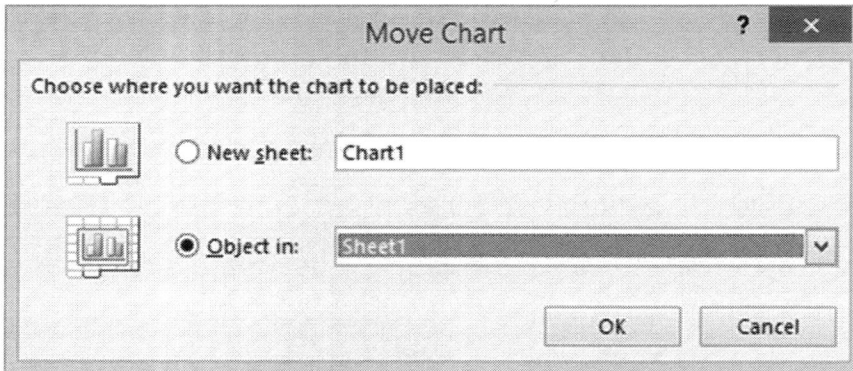

Step 4: Select New Sheet.

Step 5: Give the new worksheet a new name, if desired.

Step 6: Select OK.

Excel creates a new worksheet in the workbook (notice the tabs at the bottom). The chart has also been resized to fill the worksheet.

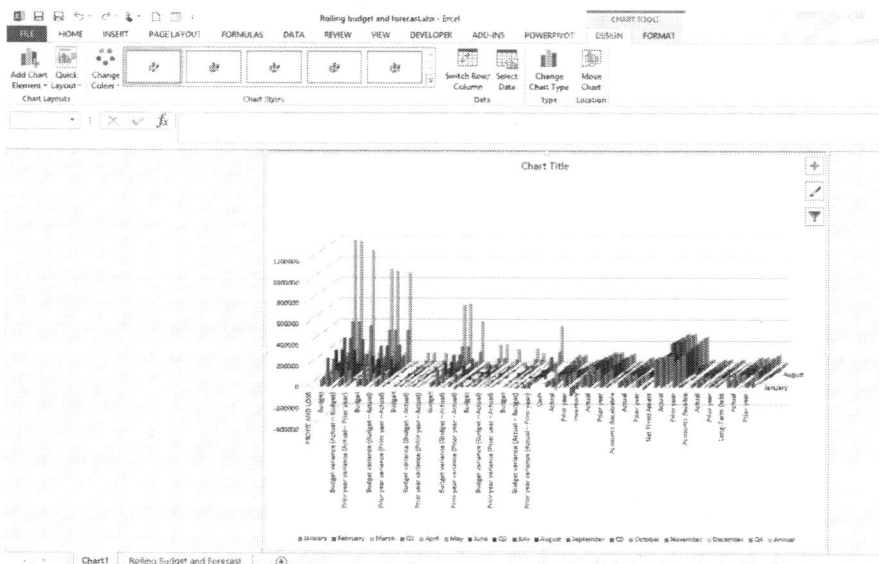

Now that you have decided what type of chart to use, it is time to make your data shine. Those little icons to the right of your chart are new to Excel 2016 and will help you add chart elements, change the style and color scheme, and use data filters. You will also learn about adding and working with data labels.

Using Chart Elements

To add a chart element, use the following procedure.

Step 1: Select the + sign on the right side of your chart.

Step 2: Check the box of the element you want to add. (Or clear the box for the element you want to remove). Many of the elements include a small arrow to the right of the option. Click the arrow to see additional options.

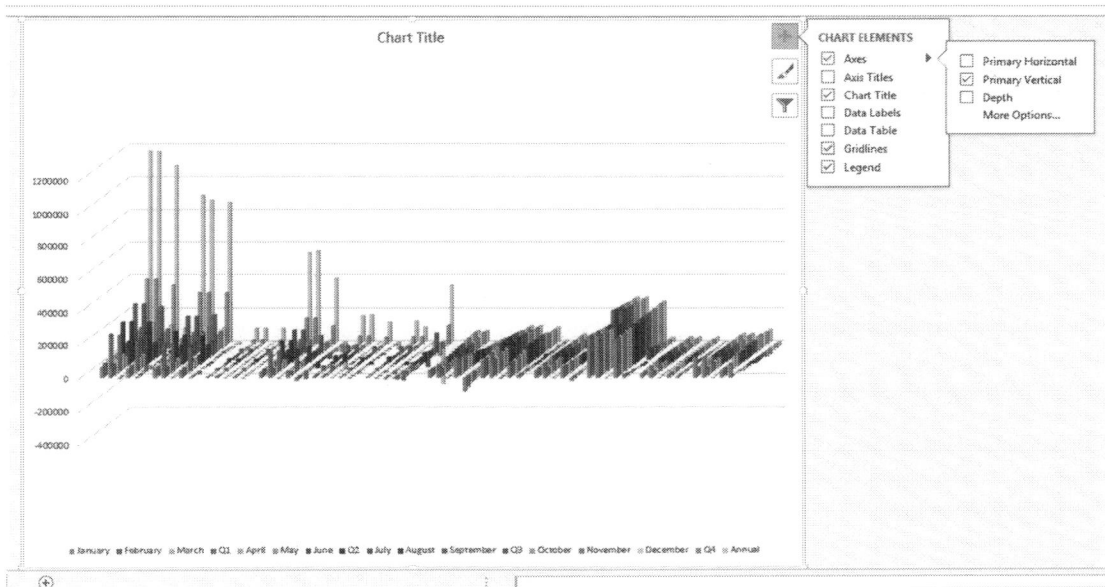

You can also use the Add Chart Element tool from the Chart Tools Design tab on the Ribbon. Pause your pointer over an option to see a preview.

To access the additional formatting options for one or more elements, use the following procedure.

Step 1: Select More Options from either Chart Elements list (from the icon next to the chart or the Ribbon).

Step 2: The Format pane opens for the selected element.

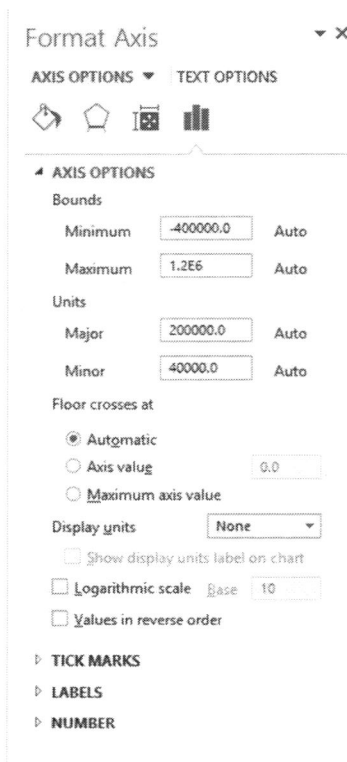

Select each of the icons to see the different options, such as Fill and Line; Shadow, Glow Soft Edges and 3-D Format; Alignment; Text Fill and Outline, and more, depending on which element you selected. You can then view the options for other elements.

Using Chart Styles and Colors

To select a new chart style, use the following procedure.

Step 1: Select the chart you want to format.

Step 2: Select the paintbrush icon to the right of your chart.

Step 3: Select the desired chart style.

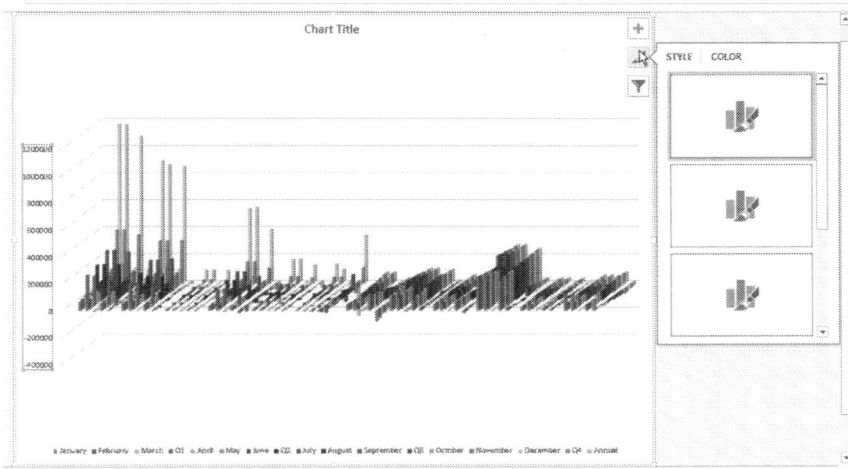

To select new chart colors, use the following procedure.

Step 1: Select the chart you want to format.

Step 2: Select the paintbrush icon to the right of your chart.

Step 3: Select the Color tab.

Step 4: Select the desired color palate.

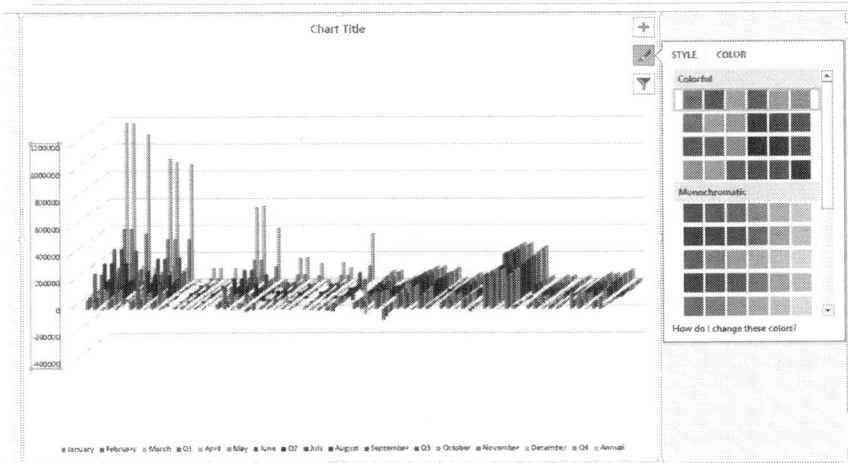

Using Chart Filters

Step 1: Select the funnel sign on the right side of your chart.

Step 2: Check the box of the value you want to add. (Or clear the box for the value you want to remove).

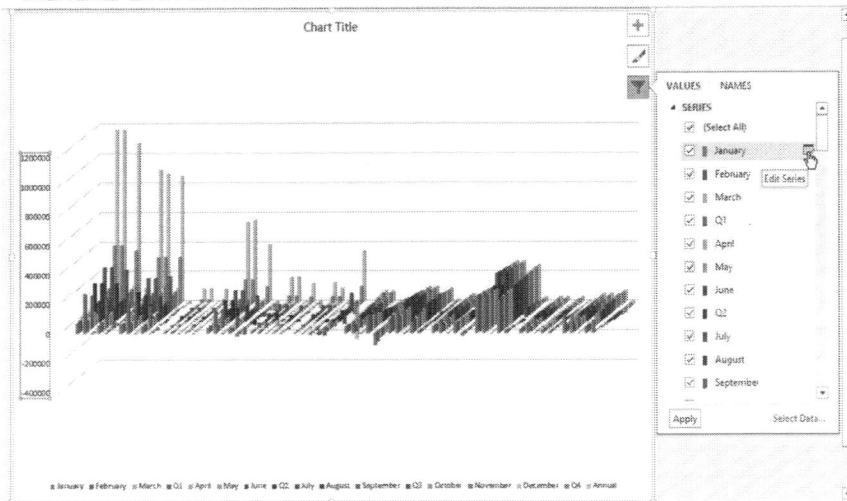

Step 3: To edit a Series name, select the worksheet icon next to the series.

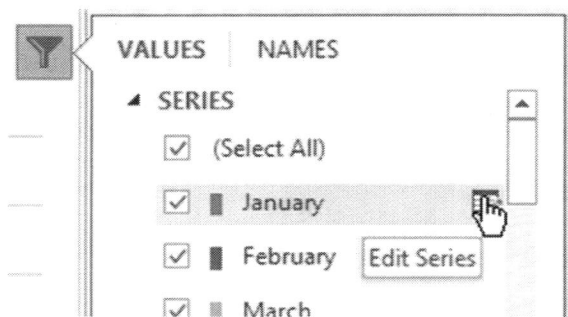

Step 4: In the Edit Series dialog box, enter the name you want to use.

Step 5: Select OK.

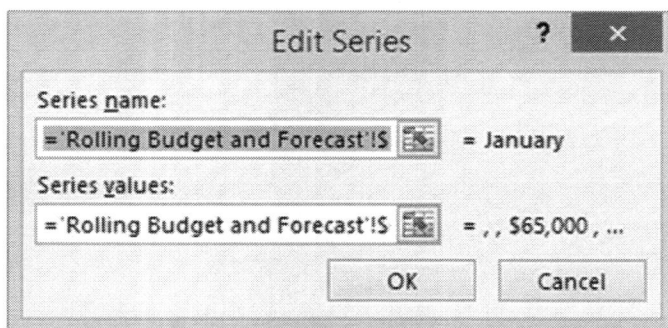

Working with Data Labels

To add data labels to a chart, use the following procedure.

116

Step 1: Select the chart that you want to label. Select the data series (such as one group of columns) or an individual data point to label, if desired.

Step 2: Select the + sign to the right of the chart.

Step 3: Check the Data Labels box and select an option from the list to the right, if desired. The options are different, based on which type of chart you have or which type of data point or series you selected.

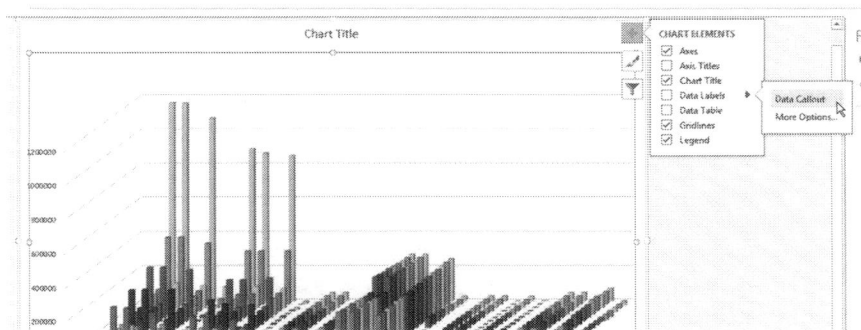

To format data labels, use the following procedure.

Step 1: Select the chart that you want to label. Select the data series (such as one group of columns) or an individual data point to label, if desired.

Step 2: Select the + sign to the right of the chart.

Step 3: Select the small arrow to the right of Data Labels.

Step 4: Select More Options.

Step 5: In the Format Data Labels pane, you can set the Fill, Border, Effects, Size & Properties, Label Options, Text Fill, Text Effects, and Text Box options. Review each of these views of the Format Data Labels pane. Note that each set of options can be expanded or condensed using the small arrow to the left of the title.

Line & Fill

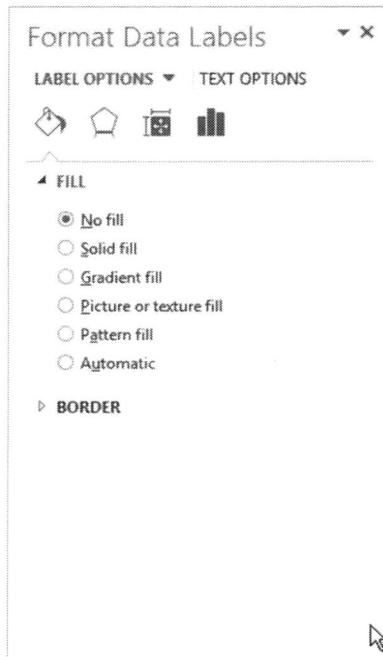

Effects

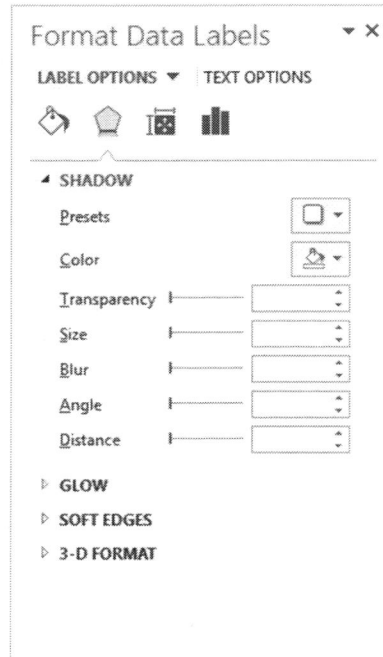

Size & Properties

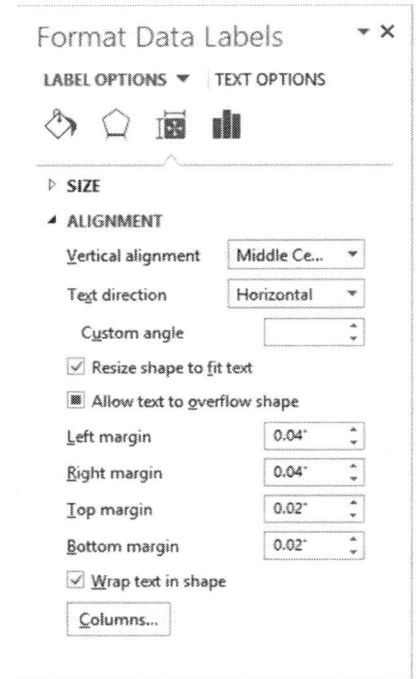

Label Options

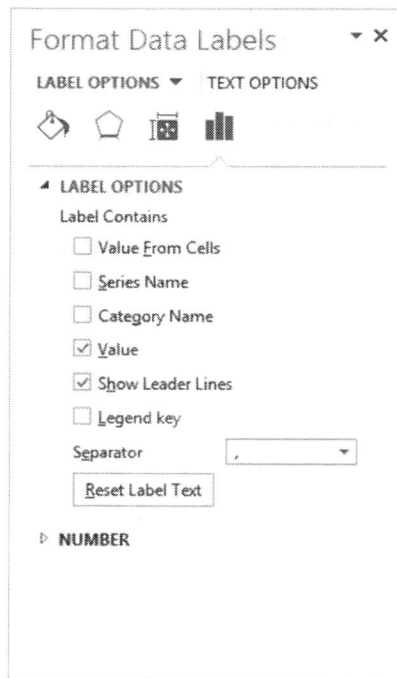

Text Fill & Outline

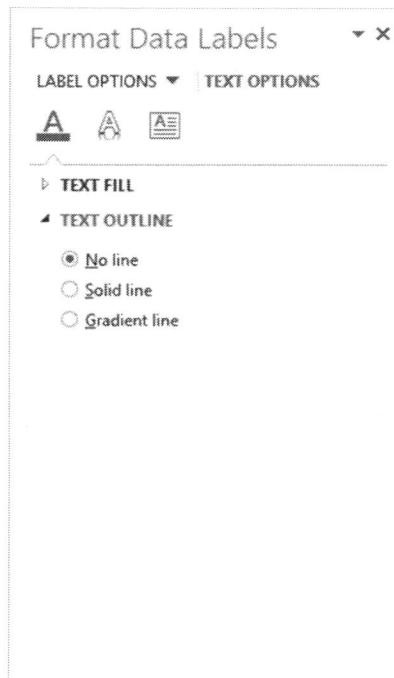

Text Effects

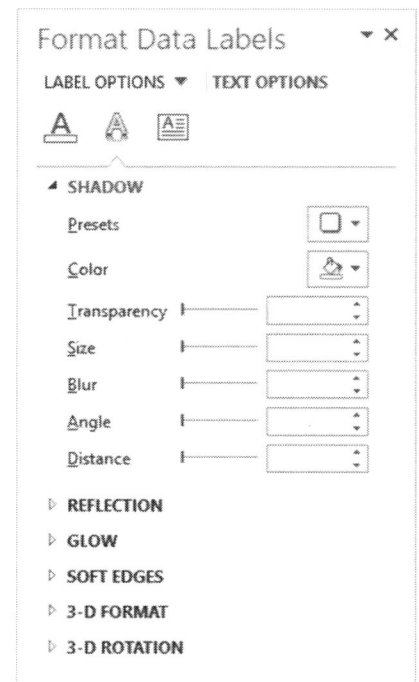

Textbox

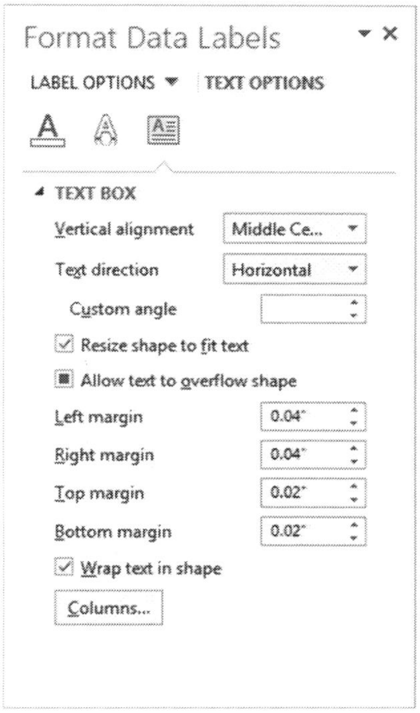

Format Data Labels

LABEL OPTIONS ▼ | TEXT OPTIONS

◢ TEXT BOX

Vertical alignment [Middle Ce... ▼]

Text direction [Horizontal ▼]

Custom angle [↕]

☑ Resize shape to fit text

▣ Allow text to overflow shape

Left margin [0.04" ↕]

Right margin [0.04" ↕]

Top margin [0.02" ↕]

Bottom margin [0.02" ↕]

☑ Wrap text in shape

[Columns...]

Chapter 15 – Creating Pivot Tables and Pivot Charts

PivotTables allow you to analyze numeric data in depth. You can use this tool to answer unanticipated questions about data. PivotTables are interactive, cross-tabulated Excel reports that summarize and analyze data. In this chapter, you will learn how to insert a Pivot Table using Excel Recommendations. You will also learn how to choose fields and group data to create different types of Pivot Tables. You will gain an understanding of the PivotTable Tools tab. You will also learn how to change the data displayed and refresh the table. You will also learn how to create a PivotChart, both from an existing PivotTable and straight from data. Finally, we will look at some real-life examples of using PivotTables and PivotCharts.

Inserting a PivotTable using Excel Recommendations

To insert a PivotTable from recommendations, use the following procedure.

Step 1: Highlight the data you want to analyze.

Step 2: Select the Insert tab from the Ribbon.

Step 3: Select Recommended PivotTables.

Excel displays the Recommended PivotTables dialog box.

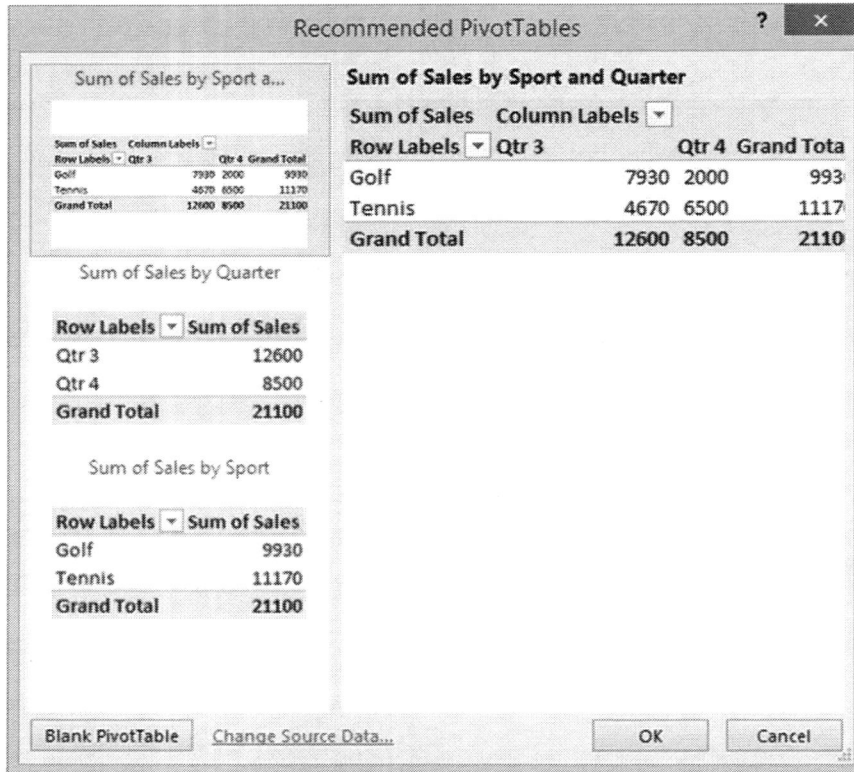

Step 4: Select the Pivot Table you want to use from the choices on the left side of the dialog box. You can see a preview on the right.

Step 5: Select OK.

Excel displays your PivotTable on a new worksheet.

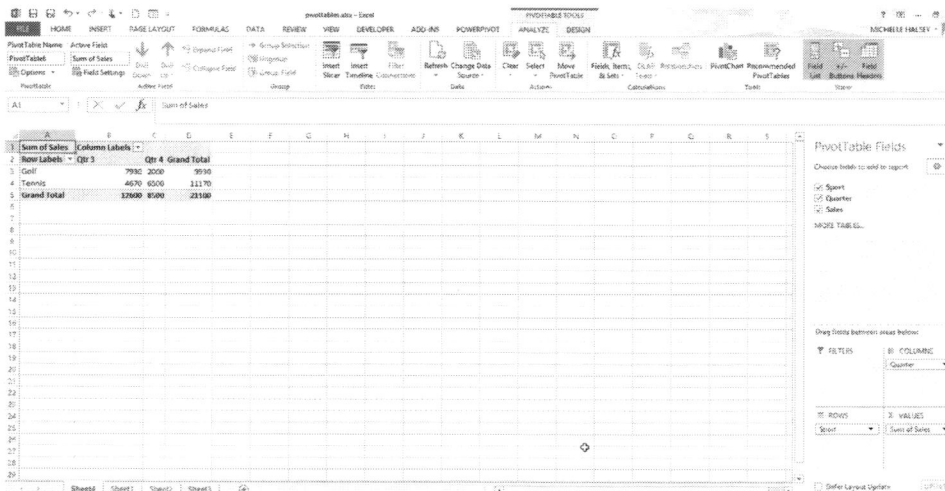

To add (or remove) fields to/from the PivotTable report, use the following procedure.

Step 1: Check the box next to a field listed in the PivotTable Fields pane to include it in the report. Clear a box to remove it. The default location where fields are added are as follows:

- Nonnumeric fields are added to the Row Labels.
- Numerical fields are added to the Values area.
- Date and time values are added to the Column Labels.

You can also apply a filter or sorting to one or more fields, by selecting the small arrow to the far right of the field name.

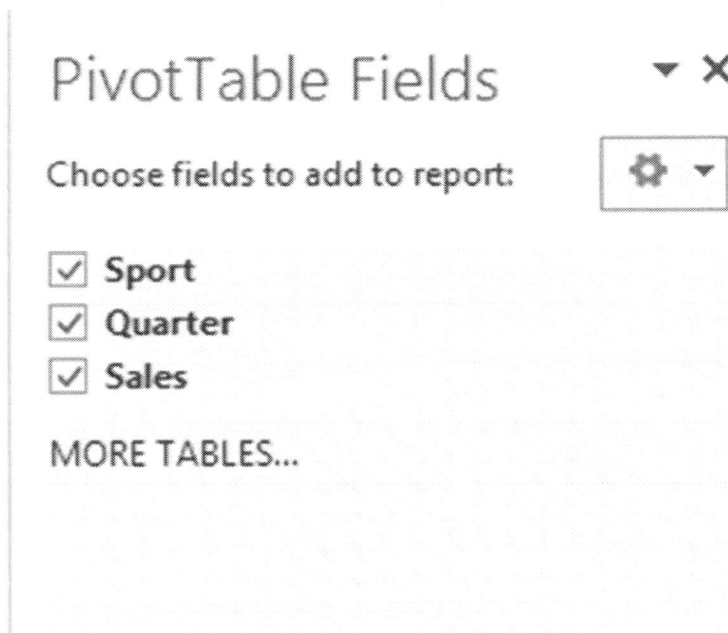

PivotTable Fields ▼ ✕

Choose fields to add to report: ⚙ ▼

- ☑ **Sport**
- ☑ **Quarter**
- ☑ **Sales**

MORE TABLES...

The bottom of the PivotTable Field List pane includes four areas:

- Filters
- Rows
- Columns
- Values

To group the data, use the following procedure.

Step 1: Right click on a field label in the PivotTable Field List and select one of the options from the context menu.

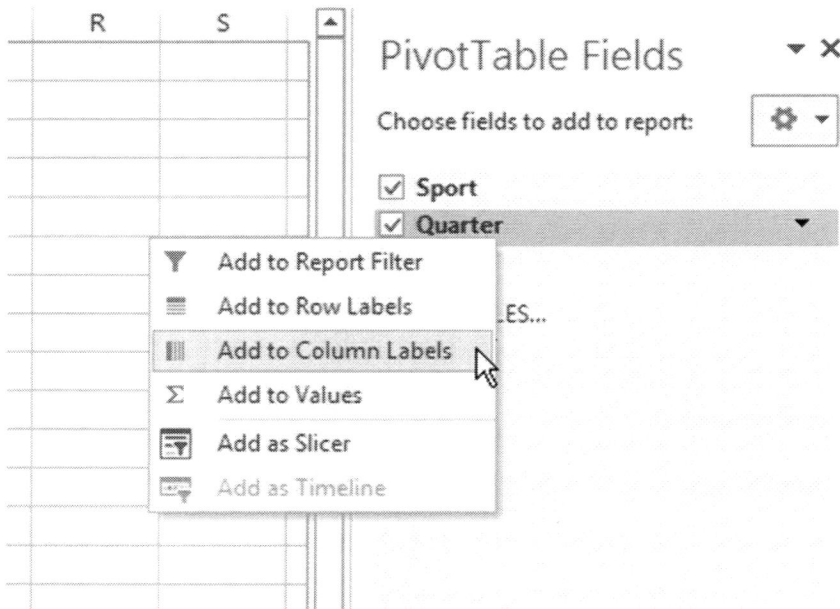

Step 2: You can also simply drag the fields from one area to another. You can even drag a field from the top portion of the pane to one of the bottom areas.

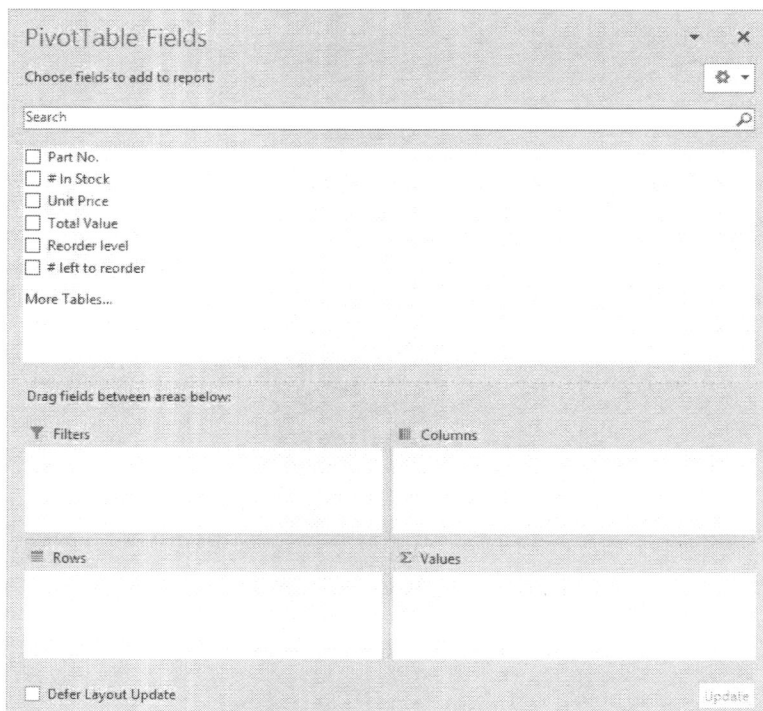

The Tools tabs for working with PivotTables.

The Analyze tab allows you to move fields, group selections, filter and work with data, perform calculations and provide tools to work with Pivot Charts in addition to the tables.

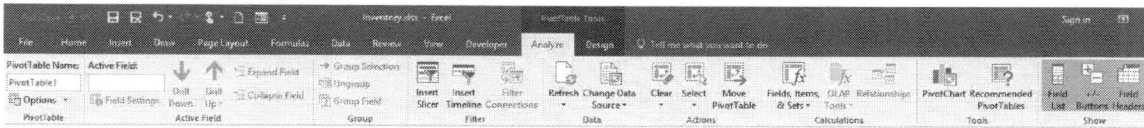

The Design tab allows you to apply predefined styles, and layout the pivot tables for optimum visualization of data.

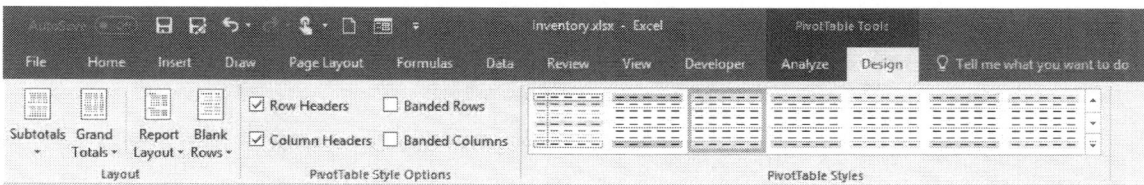

Changing the Data Displayed and Refreshing the PivotTable

To refresh the PivotTable after making a change to the data, use the following procedure.

Step 1: Return to the worksheet containing the PivotTable.

Step 2: Click somewhere on the PivotTable.

Step 3: Select the Options tab from the Ribbon.

Step 4: Select Refresh.

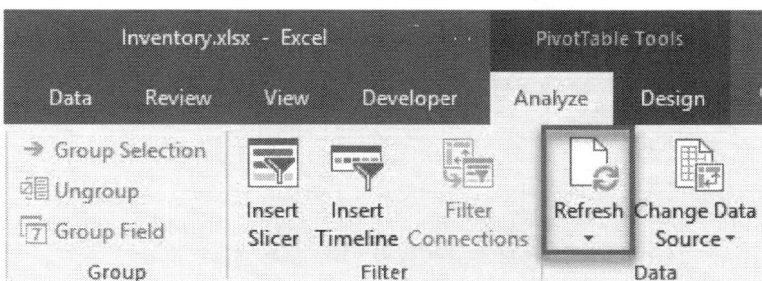

To change the data source, use the following procedure.

Step 1: Select the PivotTable Tools Analyze tab from the Ribbon.

Step 2: Select Change Data Source.

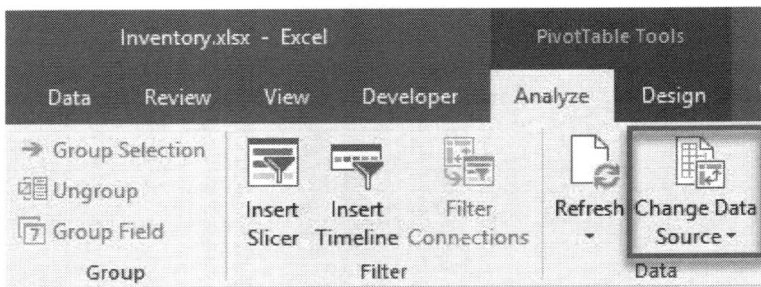

Excel returns to the worksheet of the source data and highlights the current data source. It also displays the Change PivotTable Data Source dialog box.

Step 3: Highlight the new data area on the worksheet.

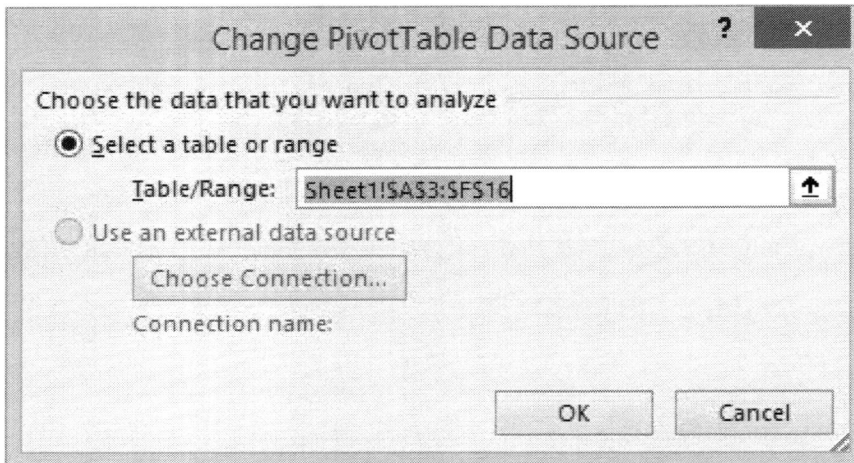

Step 4: Select OK.

Step 5: Excel opens the new PivotTable. Select and group the fields as desired.

Creating a Pivot Chart from a Pivot Table or Data

To add a PivotChart from a PivotTable, use the following procedure.

Step 1: Click anywhere in the PivotTable for which you want to add a chart.

Step 2: Select the PivotTable Tools Analyze tab from the Ribbon.

Step 3: Select PivotChart.

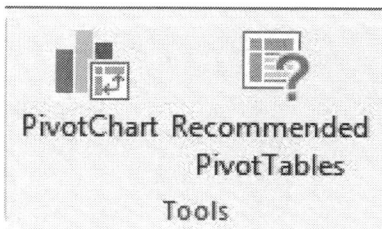

Excel displays the Insert Chart dialog box.

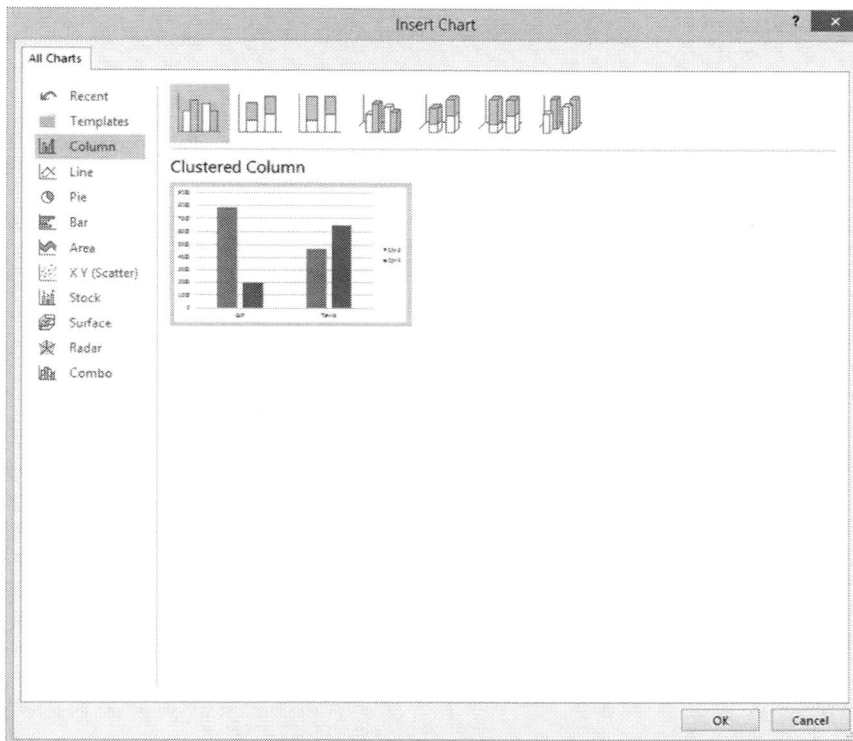

Step 4: Select the desired type of chart and select OK.

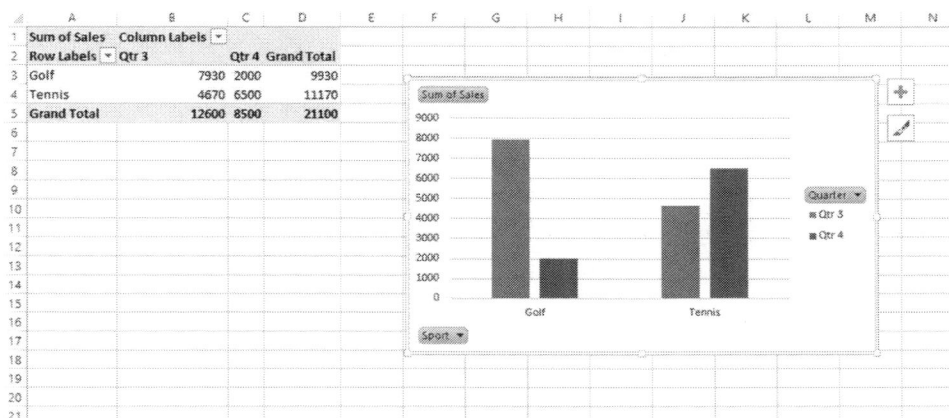

To insert a PivotChart from data, use the following procedure.

Step 1: Place your cursor somewhere in the data you want to analyze.

Step 2: Select the Insert tab from the Ribbon.

Step 3: Select PivotChart.

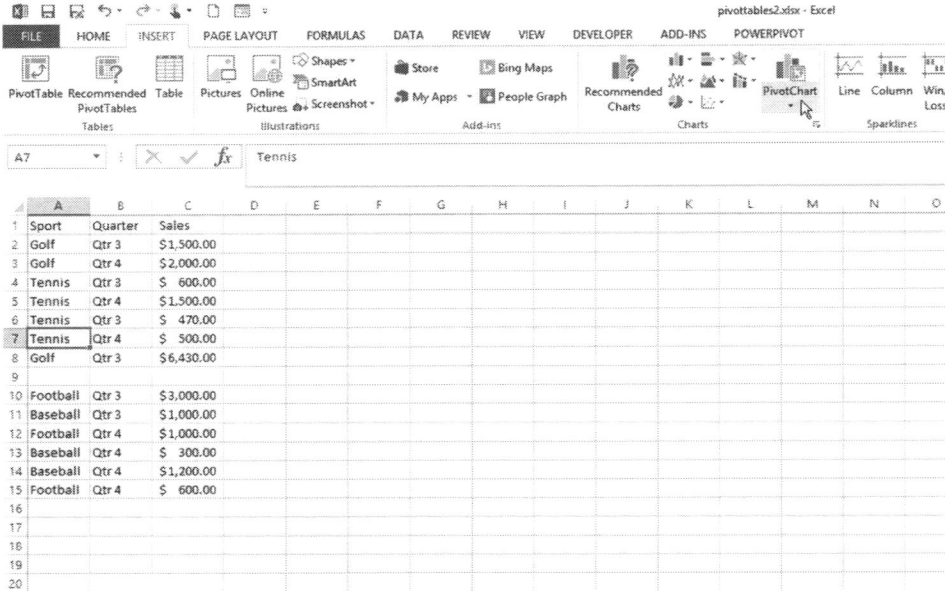

Excel displays the Create PivotChart dialog box.

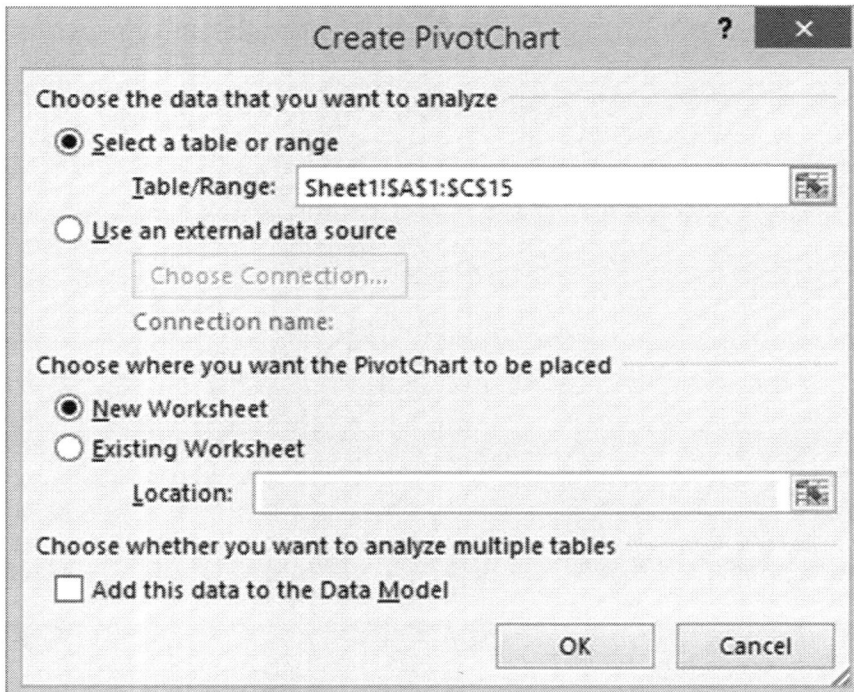

Step 4: Excel automatically provides a range of cells based on your selection. You can change the table or range if desired.

Step 5: Select a location for the PivotChart. You can have Excel create a new worksheet or select one of the existing sheets.

Step 6: Select OK.

Excel displays the blank PivotChart and the Fields pane for you to begin choosing your fields and grouping data.

Step 7: Add fields to view the chart.

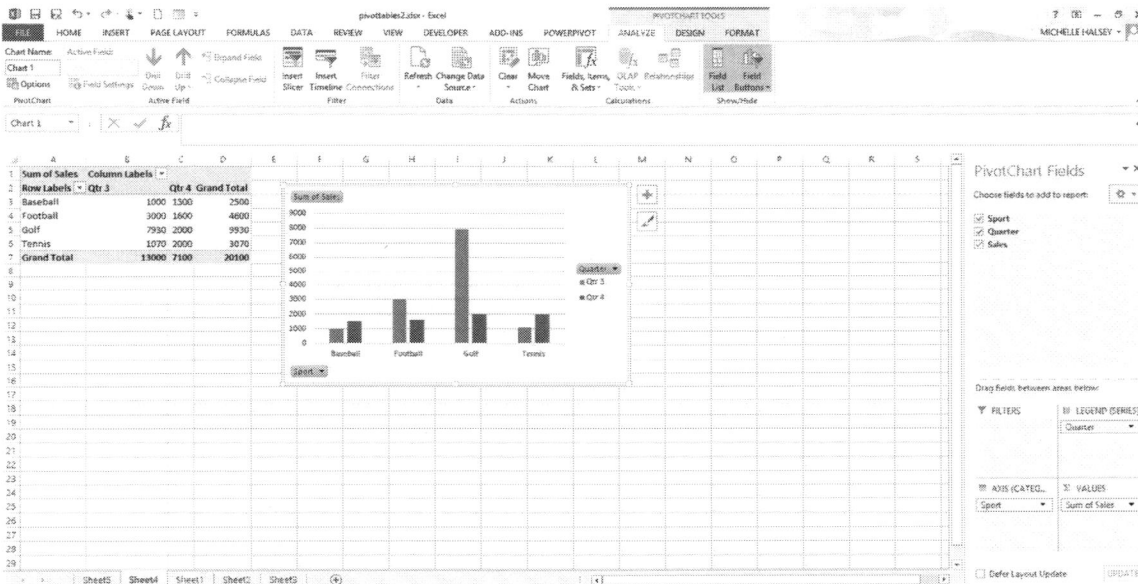

Some Real-life Examples

Find out the sum of payments from each client.

Find out the sum of payments by month.

If you create a PivotTable that includes the Salesperson and the Order Amount, you get this.

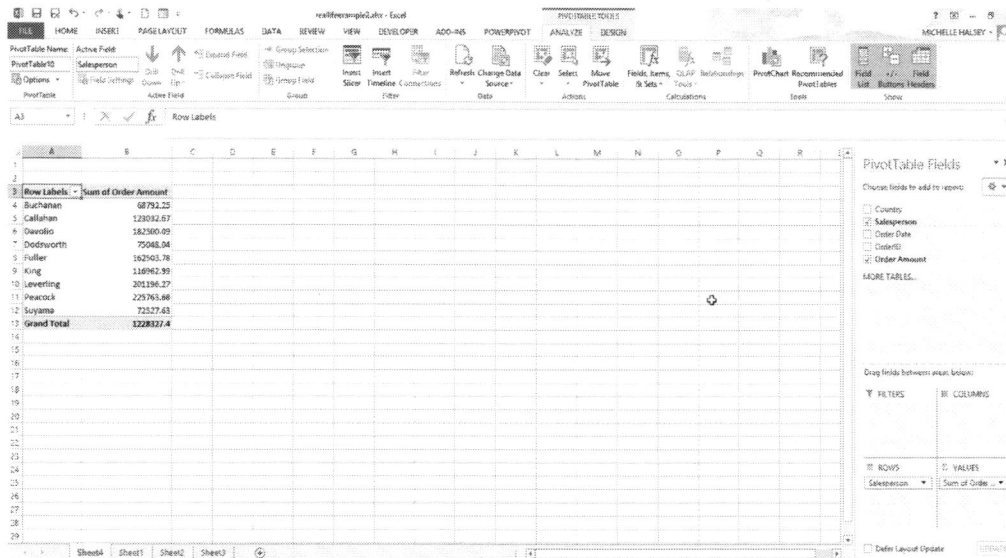

Who are the top ten salesmen?

Click the arrow next to Row labels and select More Sort Options.

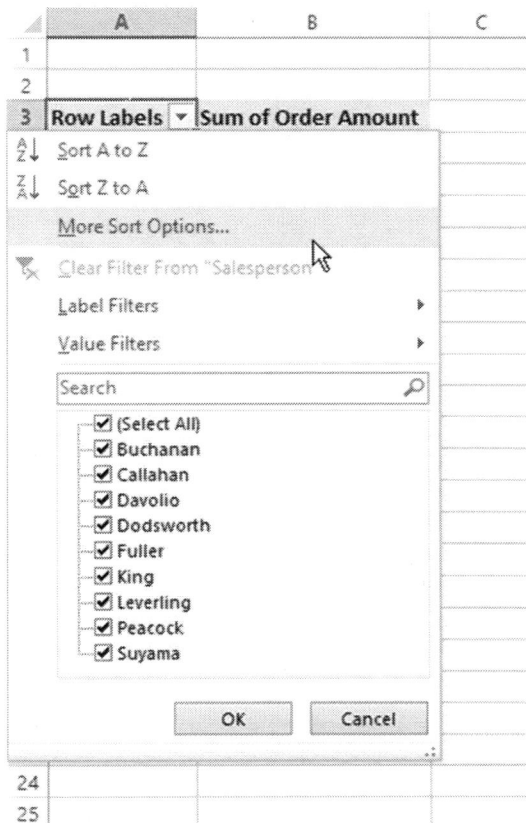

In the Sort dialog box, select Ascending (A to Z) by Sum of Order Amount and select OK.

Sort (Salesperson)   ?   ×

Sort options

- ◉ Manual (you can drag items to rearrange them)
- ○ Ascending (A to Z) by:

  Salesperson   ⌄

- ○ Descending (Z to A) by:

  Salesperson   ⌄

Summary

Drag items of the Salesperson field to display them in any order

| More Options... | OK | Cancel |

A macro is a set of recorded computer instructions. These instructions are associated with a shortcut key or macro name that makes it easy to tell your computer to run that set of instructions. This chapter will explain how to save time with macros. You will learn how to display the Developer tab, which contains the tools you will need to record macros. You will learn how to record and run macros. This chapter also explains macro security levels to avoid allowing malicious content to damage your computer with macros. Finally, you will learn how to customize and change the Quick Access Toolbar so that you have instant access to your favorite macros.

Displaying the Developer Tab

To display the Develop tab, use the following procedure.

Step 1: Select the File tab from the Ribbon.

Step 2: Select Options.

Step 3: Select Customize the Ribbon.

Step 4: In the Customize the Ribbon list on the right, check the Developer box.

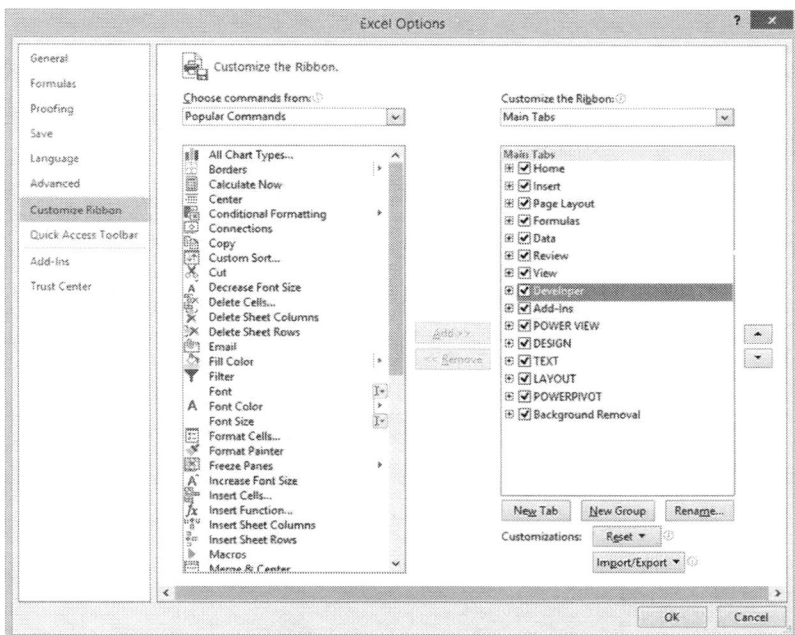

Step 5: Select OK.

The Developer tab.

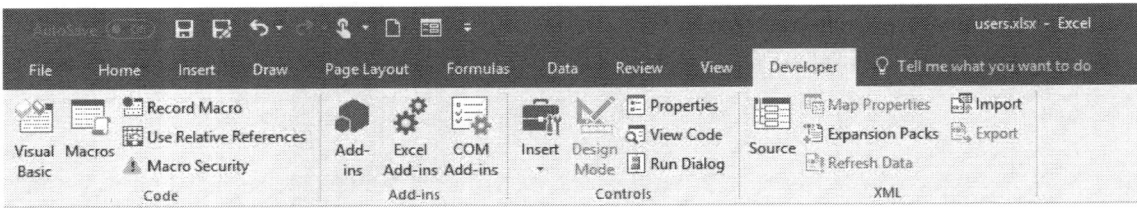

Recording and Running Macros

To record a macro, use the following procedure. In this example, we will sum the column and add formatting to the numbers.

Step 1: Select the Developer tab from the Ribbon.

Step 2: Select Use Relative References.

Step 3: Select Record Macro.

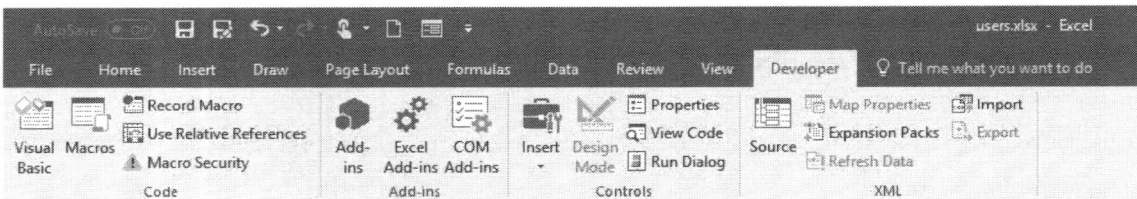

Step 4: In the Record Macro dialog box, give your macro a name.

Step 5: To make the macro available to other worksheets, select Personal Macro Workbook from the Store Macro In drop down list.

Step 6: Select OK to begin recording.

Step 7: Perform the actions you want to record. In this example, we inserted a Sum and then formatted the total with a currency formatting and added bold face formatting.

Step 8: Select the Developer tab.

Step 9: Select Stop Recording.

To run a macro, use the following example.

Step 1: Place your cursor in the cell where you want to perform the macro.

Step 2: Select the Developer tab.

Step 3: Select Macros.

Step 4: In the Macro dialog box, select your macro name from the list.

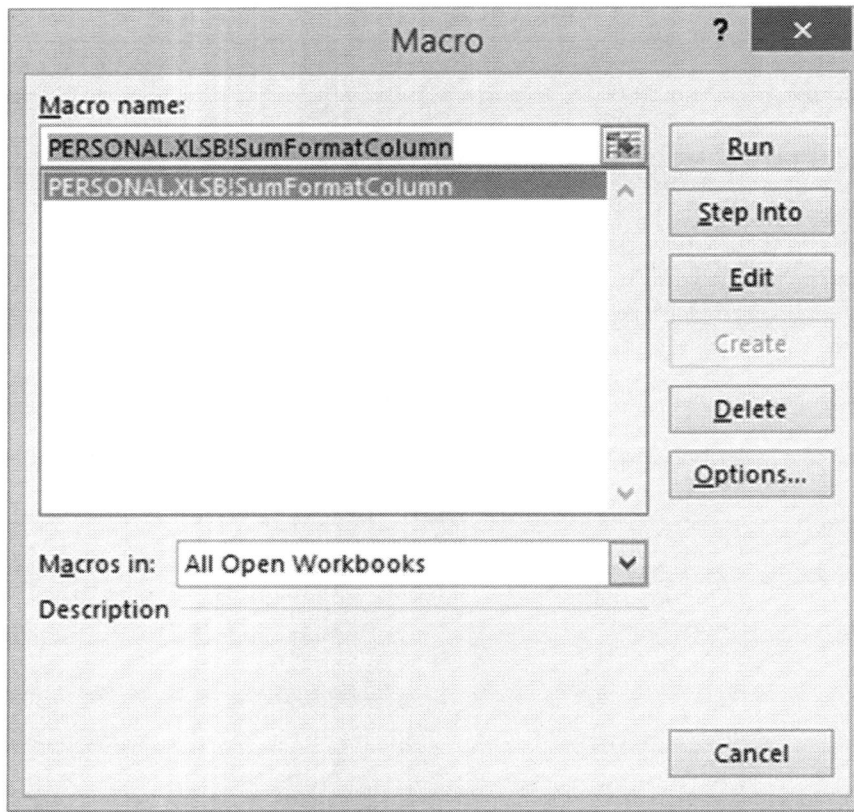

Step 5: Select Run.

Note that when you close Excel, you will get the following warning message.

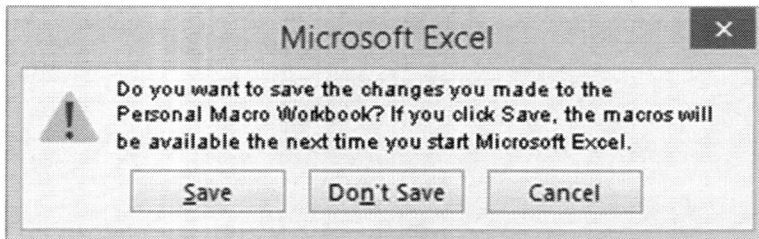

Select Save to keep the macro and make it available to other workbooks.

Changing the Security Level

To change the macro security, use the following procedure.

Step 1: Select the Developer tab.

Step 2: Select Macro Security.

Step 3: Select one of the following options:

Disable all macros without notification – this option only runs macros in documents in trusted locations.

Disable all macros with notification – this option disables macros that are not in trusted locations, but it provides notification, so that you can choose to enable those macros on a case by case basis.

Disable all macros except digitally signed macros – this option allows not only macros in trusted locations, but also macros that are digitally signed by a trusted publisher. Other macros are disabled with notification to allow you to choose to enable those macros on a case by case basis.

Enable all macros – this option allows all macros to run, which is potentially dangerous since virus authors often use macros to distribute malicious code on computers. Microsoft does not advise using this setting.

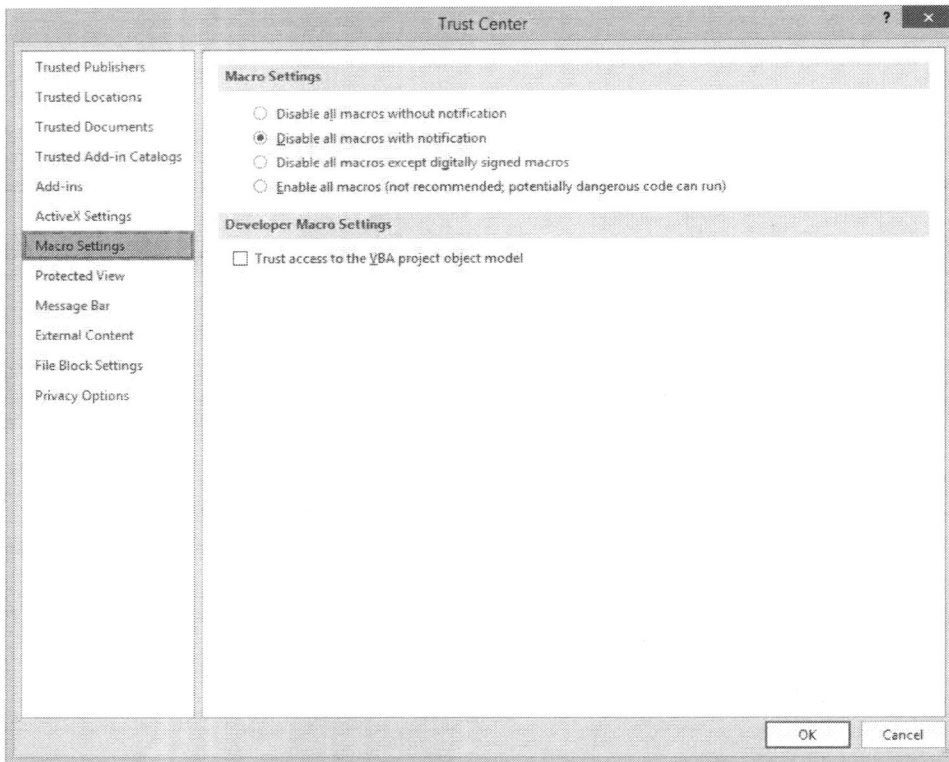

Step 4: Check the Trust Access to the VBA Project Object Model box only if you are a developer. This security option makes it more difficult for unauthorized programs to build code that self-replicates.

Step 5: Select OK.

Customizing and Changing the Quick Access Toolbar

To add a macro to the Quick Access Toolbar, use the following procedure.

Step 1: Select the arrow to the right of the Quick Access Toolbar.

Step 2: Select More Commands.

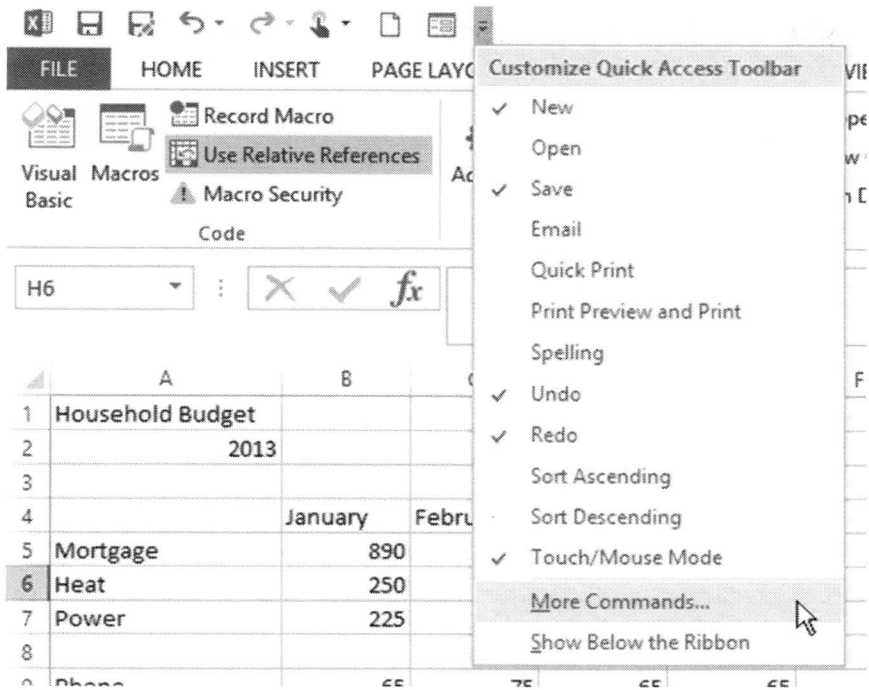

Step 3: In the Choose Command From drop down list, select Macros.

Step 4: The macro you recorded should be listed. Select it and select Add.

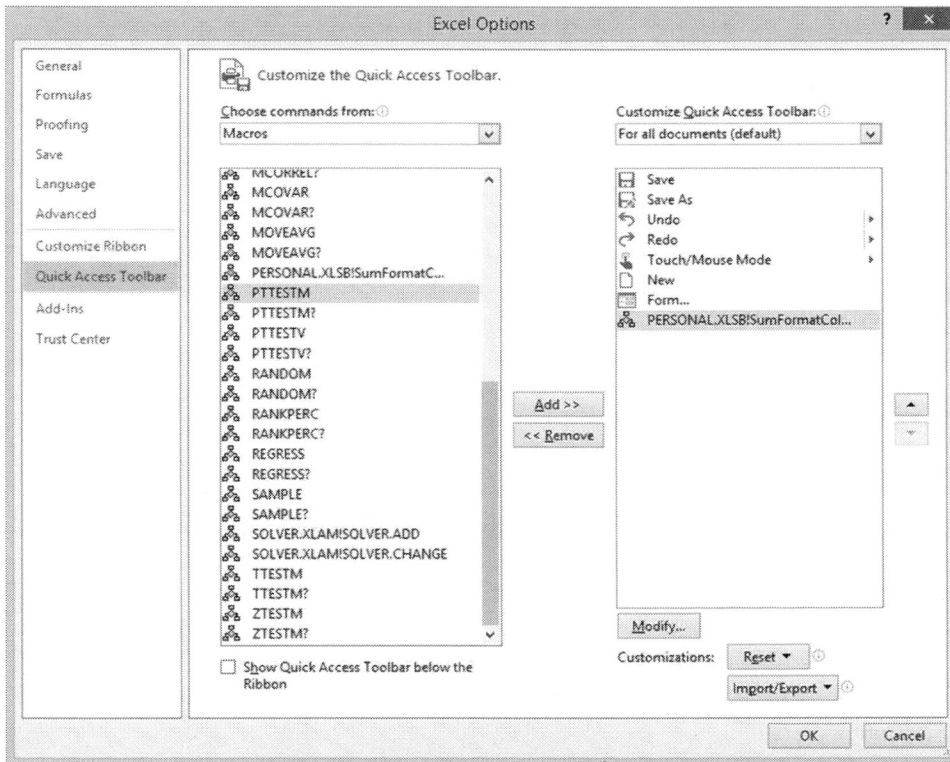

Step 5: If you would like to modify the name of the macro, select Modify.

Step 6: In the Modify Button dialog box, you can choose an icon to show in the Quick Access Toolbar. You can also modify the name.

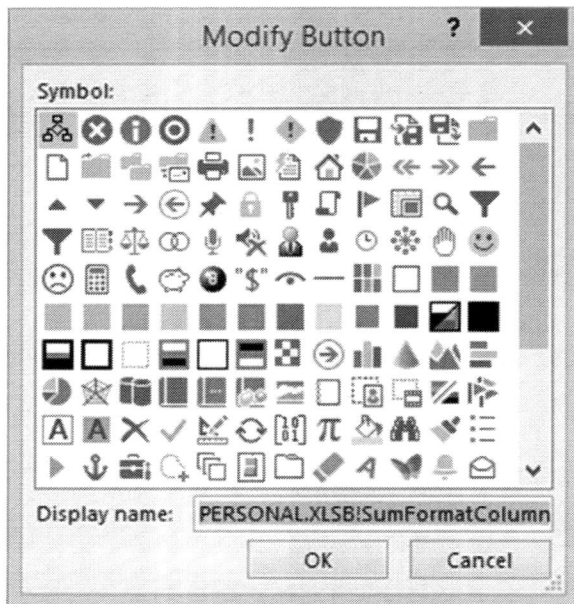

Step 7: Select OK.

Step 8: Select OK in the Excel Options window.

Chapter 17 – Solving Formula Errors

Formula errors can be very frustrating. This chapter will teach you how to prevent formula errors by using named ranges. You will gain an understanding of formula errors and learn how to use error checking. You will also learn how to use the Trace Errors commands. Finally, we will look at how to evaluate formulas.

Using Named Ranges

To name a range, use the following procedure.

Step 1: Highlight the cell references you want to name.

Step 2: Select the Formulas tab.

Step 3: Select Define Name.

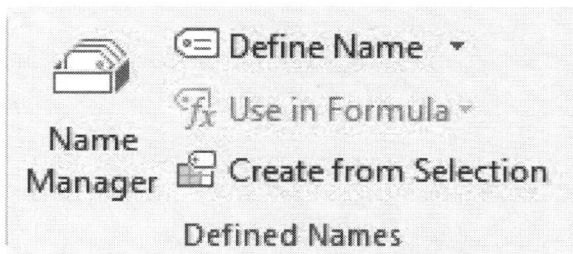

Step 4: Enter a name for the cell reference range.

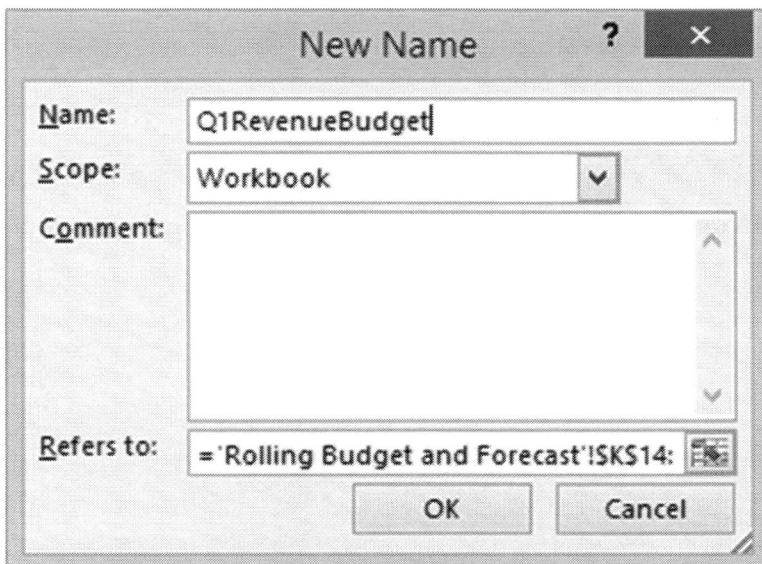

Step 5: Select a different scope for the reference, if desired, from the Scope drop down list.

Step 6: Enter a Comment, if desired,

Step 7: Change the Refers to area, if desired.

Step 8: Select OK.

To use a named range in a formula, use the following procedure.

Step 1: Begin entering your formula.

Step 2: When you are ready to enter the range, select the Formula tab.

Step 3: Select Use in Formula.

Step 4: Select the named range from the list.

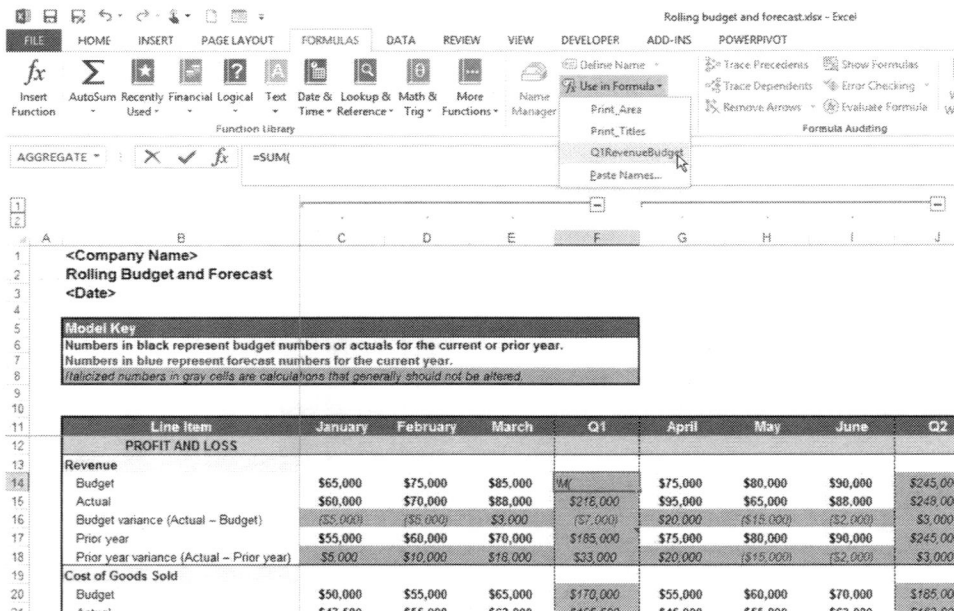

Understanding Formula Errors

Discuss the most common formula errors and how they occur.

Start every function with the equal sign (=)

Excel will display the formula contents as text or a date if you do not use the equal sign.

Match all open and close parentheses

Every parenthesis needs a pair. Parenthesis must be in the correct position for the formula to work correctly.

Use a colon to indicate a range

When working with a range of cells, you must use a colon between the first and last cell reference.

Enter all required arguments

Some functions require arguments and some do not. If the function requires arguments, make sure you have the right number.

Enter the correct type of arguments

For functions that require arguments, make sure you have the right ones.

Use the * symbol when multiplying numbers

The * symbol (asterisk) is the multiplication operator in Excel, not "x."

Use quotation marks around text in formulas

If you create a formula that includes text, enclose the text in quotation marks.

Nest no more than 64 functions

The top limit of nested functions, or functions within a function, is 64.

Enclose other sheet names in single quotation marks

If your worksheet names contain non-alphabetical characters, you must enclose the sheet name within single quotation marks when using the name in a formula.

Place an exclamation point (!) after a worksheet name when you refer to it in a formula

If you are using a worksheet name in a formula, the name must be followed by an exclamation point.

Include the path to external workbooks

If you are referencing cells from another workbook, make sure the formula includes both the workbook name and the path to the workbook.

Enter numbers without formatting

Excel treats commas as separator characters. Format the formula result after you enter the numbers in the formula.

Avoid dividing by zero

If you divide a cell by another that is zero or no value can result in a #DIV/0! Error.

Using Error Checking

The Error Checking dialog box, use the following procedure.

Step 1: From anywhere on the worksheet, select the Formulas tab.

Step 2: Select Error Checking.

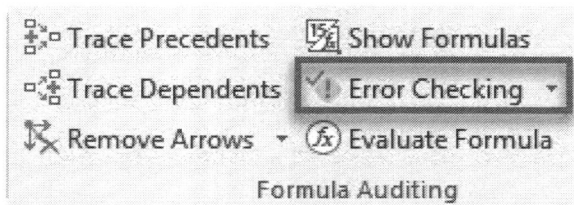

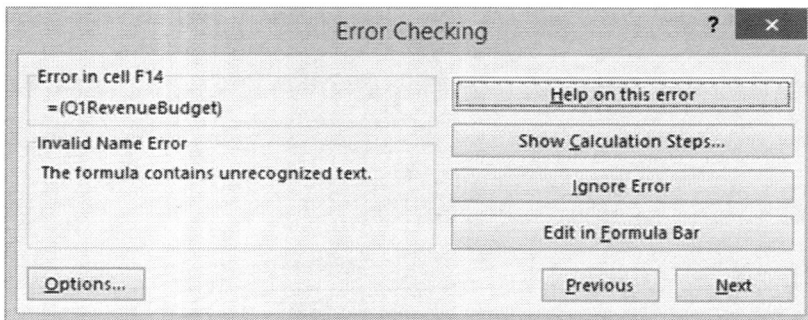

The Error Checking dialog box displays the formula as written in the cells. It explains why the formula contains an error.

- Help on this Error – opens the Excel help files directly to an article related to the type of error Excel detected.

- Show Calculation steps – opens the Evaluate Formula dialog box (discussed later in this chapter).

- Ignore Error – allows you to keep the error and removes the green triangle from the cell.

- Edit in Formula Bar – moves your cursor to the Formula bar to allow you to correct the formula.

- Options – opens the Options window to allow you to adjust the error checking options.

- Resume – restarts the Error Checking if you have switched to another task.

- Previous – returns to the previous error.

- Next – moves to the next error.

The Excel Options dialog box for Formulas.

You can open the Formulas options from the Error Checking dialog box or the Trace Errors commands next to an error cell.

You can also open the Options dialog box selecting the File tab from the Ribbon. Then select Options. Select Formulas.

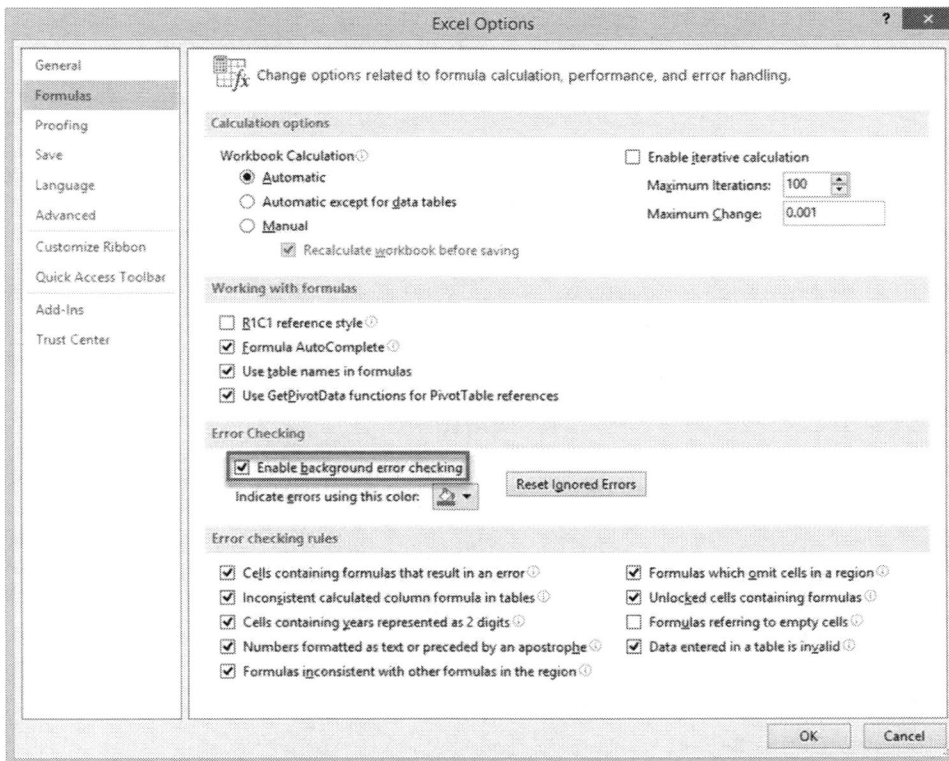

Under Error Checking, you can turn on Background error checking by checking the box.

You can change the color of the triangle displayed in cells where Excel has detected a formula error.

Select the Reset Ignored Errors to re-enable Excel to help you with any errors that you have previously ignored.

In the Error Checking Rules area, you can check or clear the following checkboxes:

- Cells containing Formulas that result in an error – When checked, Excel checks for formulas that do not use expected syntax, arguments, or data types.

- Inconsistent calculated column formula in tables – When checked, Excel checks for inconsistencies in calculated columns, such as when you enter data other than a formula in a column that has all calculated cells.

- Cells containing years represented in 2 digits – When checked, Excel will create an error if you enter a date with a year represented as two digits.

- Numbers formatted in text or preceded by an apostrophe – When checked, Excel will create an error if you enter or import numbers preceded by an apostrophe or text.

- Formulas inconsistent with other formulas in the region – When checked, Excel looks for formulas that are different from formulas near it. Often these formulas should be the same, except for the cell references used.

- Formulas which omit cells in a region – When checked, Excel compares the reference in a formula against the actual range of cells adjacent to it.

- Unlocked cells containing formulas – Formulas are locked for protection by default and must be unlocked before editing. If you have unlocked cells with formulas, Excel marks it as an error when this box is checked.

- Formulas referring to empty cells – When checked, Excel creates an error if a formula includes a reference to an empty cell.

- Data entered in a table is invalid – When checked, Excel creates an error if there is a validation error in a table.

The Trace Errors Commands on a cell with a formula error, use the following procedure.

Step 1: A formula with an error displays a green triangle in the upper left corner, along with an error icon next to the cell. Click on the arrow next to the icon to see the options.

- Help on this Error – opens the Excel help files directly to an article related to the type of error Excel detected.

- Show Calculation steps – opens the Evaluate Formula dialog box (discussed later in this chapter).

- Ignore Error – allows you to keep the error and removes the error icon and green triangle.

- Edit in Formula Bar – moves your cursor to the Formula bar to allow you to correct the formula.

- Error Checking Options – opens the Options window to allow you to adjust the error checking options (discussed later in this chapter).

Evaluating Formulas

To evaluate a formula, use the following procedure.

Step 1: Select the cell that contains the formula you want to evaluate.

Step 2: Select the Formulas tab.

Step 3: Select Evaluate Formula.

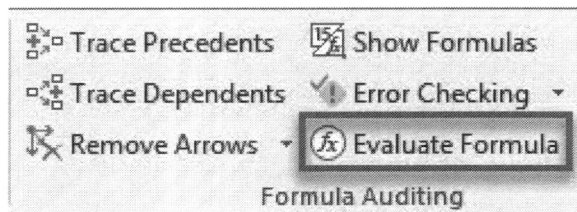

Step 4: Select Evaluate to see the results of the underlined portion of the formula.

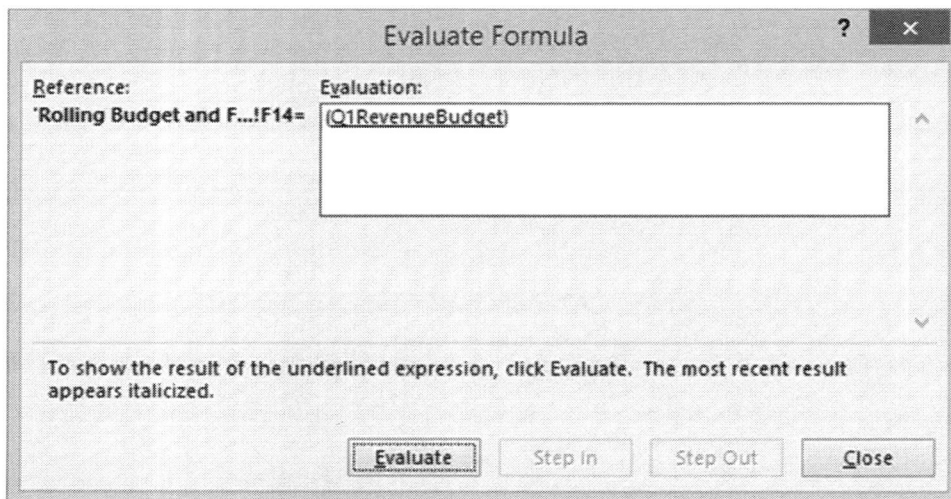

Step 5: Continue selecting Evaluate to see the results of each piece of the formula.

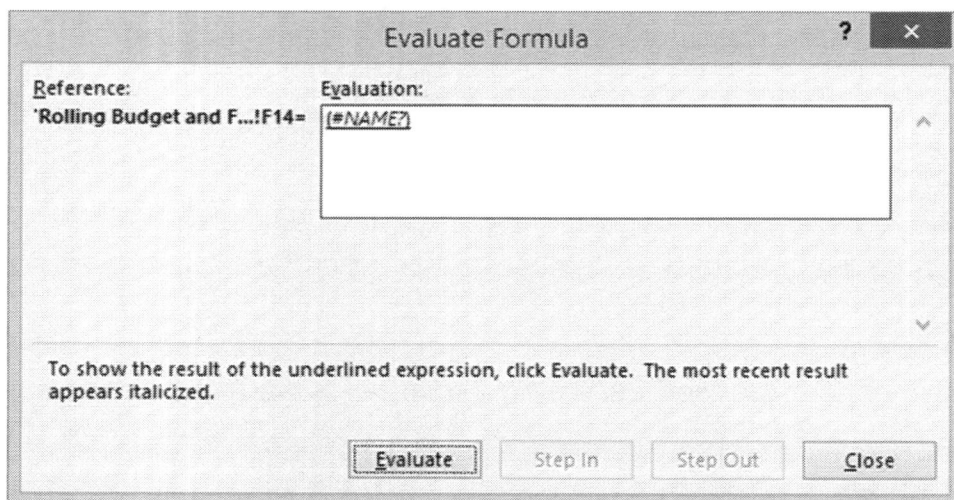

- If the underlined part of the formula is a reference to another formula, select Step In to display the other formula in the Evaluation box. Select Step Out to go back to the previous cell and formula. The Step In button is not available the second time a reference appears in the formula, or if the formula refers to a cell in a separate workbook.

- To see the evaluation again, click Restart.

Step 6: Select Close when you have finished.

Evaluate each of the formulas and discuss why the results are different, even though the same formula is used for each column of cells.

"What-if" analysis allows you to have Excel change the values in cells so that you can see how those changes affect the formulas outcomes. There are three kinds of what if analysis: goal seek, scenarios, and data tables. Goal seek allows you to find the necessary value for an unknown in a formula to obtain desired results. Scenarios allow you to view multiple different possible results for up to 32 variables. Data tables allow you to quickly calculate multiple results for one or two variables in one operation. You can view and compare the results of all the different variations together on your worksheet. This chapter introduces these tools.

Using Goal Seek

To use goal seek, use the following procedure.

Step 1: When using goal seek, one value from a formula should be left blank.

Step 2: Select the Data tab from the Ribbon.

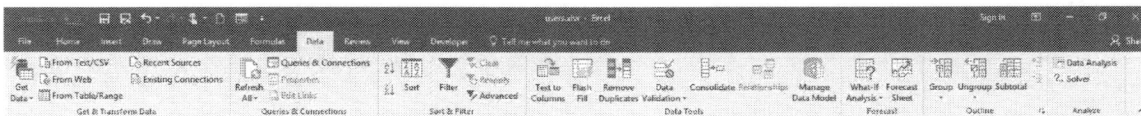

Step 3: Select What If Analysis.

Step 4: Select Goal Seek.

Excel displays the Goal Seek dialog box.

Step 1: In the Set Cell field, enter or select from the worksheet the cell that contains the formula. In the sample file, select $B$4.

Step 2: In the To Value field, enter the formula result you want. For example, in the sample file, you may want the resulting payment of $900. You would enter -900 because it is a payment.

Step 3: In the By Changing Cell field, enter or select the reference for the cell that contains the value you do not know. In the sample file, this is $B$3.

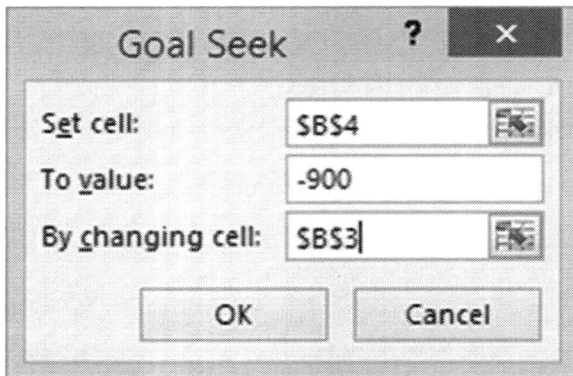

Goal Seek	?	×
Set cell:	$B$4	
To value:	-900	
By changing cell:	$B$3	
	OK	Cancel

Step 4: Select OK.

Excel displays the Goal Seek Status dialog box. Select OK to close it.

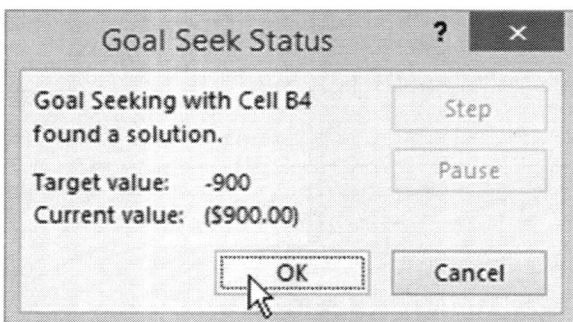

Goal Seek Status	?	×
Goal Seeking with Cell B4 found a solution.	Step	
	Pause	
Target value: -900		
Current value: ($900.00)		
	OK	Cancel

▲	A	B
1	Loan Amount	100000
2	Term in Months	180
3	Interest Rate	0.07021
4	Payment	($900.00)

You may need to reformat the cell with the new answer to view the answer in the preferred format.

Using the Scenario Manager

To add a scenario, use the following procedure.

Step 1: Select the Data tab from the Ribbon.

Step 2: Select What If Analysis.

Step 3: Select Scenario Manager.

Step 4: In the Scenario Manager dialog box, select Add to create a new scenario.

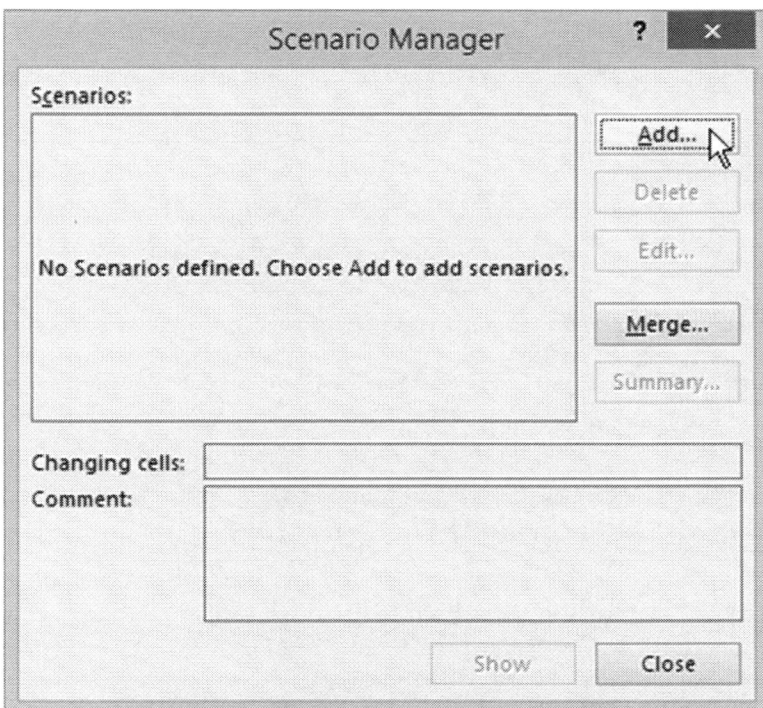

Step 5: In the Add Scenario dialog box, enter a Scenario Name.

Step 6: In the Changing Cells field, enter (or select from the worksheet) the multiple cells of changing values in the first scenario. Press the CTRL key while selecting each value.

Step 7: Enter a Comment, if desired.

Step 8: Protect the scenario by checking the Prevent changes and/or the Hide boxes.

Step 9: Select OK.

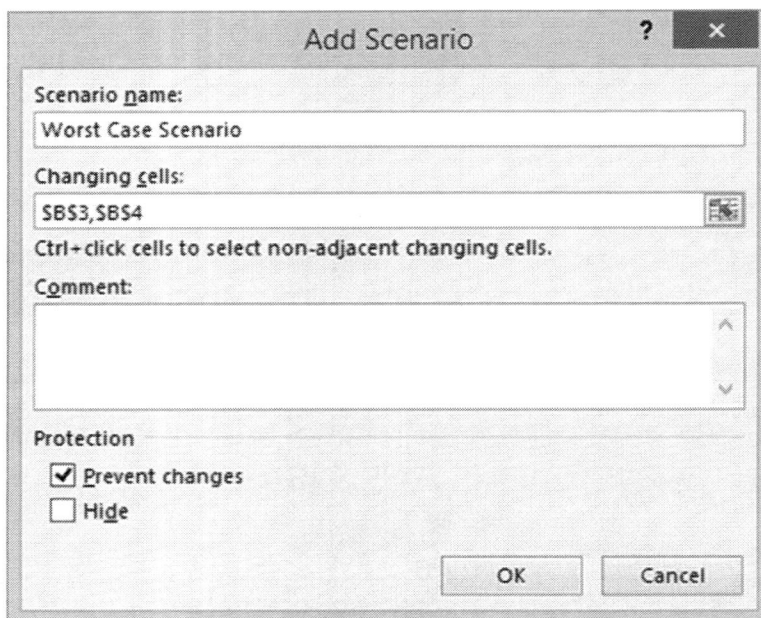

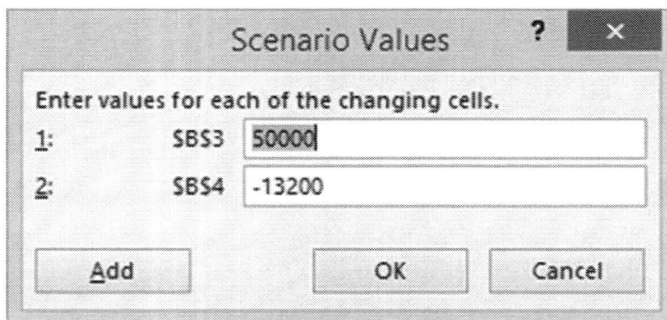

Add Scenario	?	×
Scenario name:		
Worst Case Scenario		
Changing cells:		
$BS3,$BS4		
Ctrl+click cells to select non-adjacent changing cells.		
Comment:		
Protection		
☑ Prevent changes		
☐ Hide		
	OK	Cancel

Scenario Values	?	×
Enter values for each of the changing cells.		
1: $BS3	50000	
2: $BS4	-13200	
Add	OK	Cancel

Step 10: The Scenario values dialog box shows the values you selected.

- For the original scenario, keep the values Excel displays.
- For each subsequent scenario, enter the new values.

Step 11: Select Add to create another set of values. If you have finished adding all the possibilities, select OK to return to the Scenario Manager.

Step 12: Repeat steps 4 through 10 to create another scenario.

Step 13: On the Scenario Manager dialog box, you can select a scenario name and select Show to see the results. The contents of the cells change, depending on which scenario you select and show. To view a report, select Summary.

Excel displays the Scenario Summary dialog box.

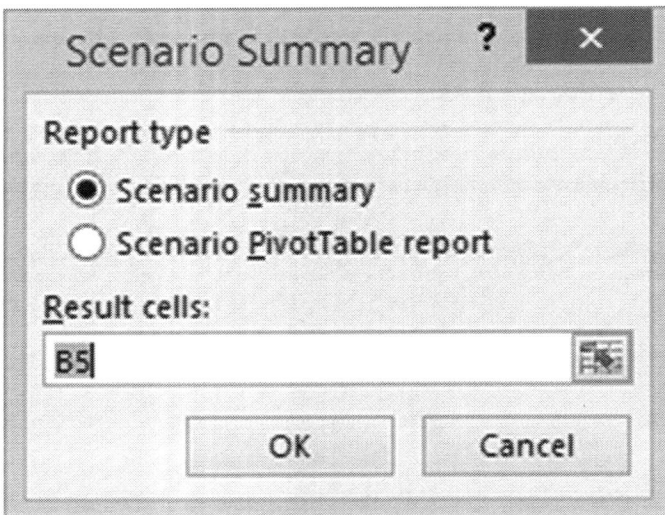

Step 14: Indicate whether Excel should display the Scenario Summary or a Scenario PivotTable Report.

Step 15: Select the cell that contains the results you want to compare (or the formula cell).

Step 16: Select OK.

Excel displays your results in the selected format.

To set up a one-input data table, use the following procedure.

Step 1: Enter the known values that the formula will use in evaluating the variable values.

Step 2: Enter the list of values you want to use for the input cell for the formula either down one column or across one row. If you are entering the values in a column, as shown below, leave the column to the right empty. Also leave additional rows below the values empty. If you are entering the values in a row, leave the rows below the values empty. Also leave a few columns to the right empty.

	A	B	C	D
1	Mortgage Loan Analysis			Payments
2	Down Payment	None		
3	Interest Rate		9.00%	
4	Term(months)	360	9.25%	
5	Loan Amount	80000	9.50%	
6				
7				
8				
9				

Step 3: If you have entered your data in columns, enter the formula one cell above and one cell to the right of the list of data values. You can enter additional formulas in the cells to the right of this cell to evaluate how the data values affect other formulas. If you have entered your data in rows, enter the formula one column to the left of the first value and one cell below the row of values.

Step 4: Select the data table values and the formula. In this example, the range is C2:D5.

Step 5: Select the Data tab from the Ribbon.

Step 6: Select What If Analysis.

Step 7: Select Data Table.

Step 8: Select the input cell in the formula. In a one-input data table, you will only have one input. In this example, the cell B3 is the Column Input cell.

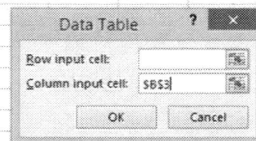

Step 9: Select OK.

For each possible value for the variable listed in the data table, Excel displays the results.

158

	A	B	C	D
1	Mortgage Loan Analysis			Payments
2	Down Payment	None		$222.22
3	Interest Rate		9.00%	643.6980936
4	Term (Months)	360	9.25%	658.1403404
5	Loan Amount	80000	9.50%	672.6833657
6				

You may want to format the cells to show the results with the desired formatting (such as currency in this example).

Using a Two Input Data Table

To set up a two-input data table, use the following procedure.

Step 1: Enter the known values that the formula will use in evaluating the variable values. In this example, using the previous lesson's workbook, delete the numbers except for the Loan Amount.

Step 2: Enter the formula. In this example, it should be entered in cell C2.

AGGREGATE ▾	:	✕ ✓ *fx*	=PMT(B3/12,B4,-B5)

	A	B	C	D	E
1	Mortgage Loan Analysis				Payments
2	Down Payment	None		=PMT(B3/12,B4,-B5)	
3	Interest Rate		9.00%		
4	Term (Months)	360	9.25%		
5	Loan Amount	80000	9.50%		
6					
7					
8					

Step 3: Enter the list of values for the first input cell for the formula down one column under the formula. In this example, the unknown interest rate is the first input cell.

	A	B	C	D
1	Mortgage Loan Analysis			Payments
2	Down Payment	None	#NUM!	
3	Interest Rate		9.00%	
4	Term (Months)		9.25%	
5	Loan Amount	80000	9.50%	
6				

Step 4: Enter the list of values for the second input cell for the formula across in one row next to the formula. In this example, the unknown term is the second input cell.

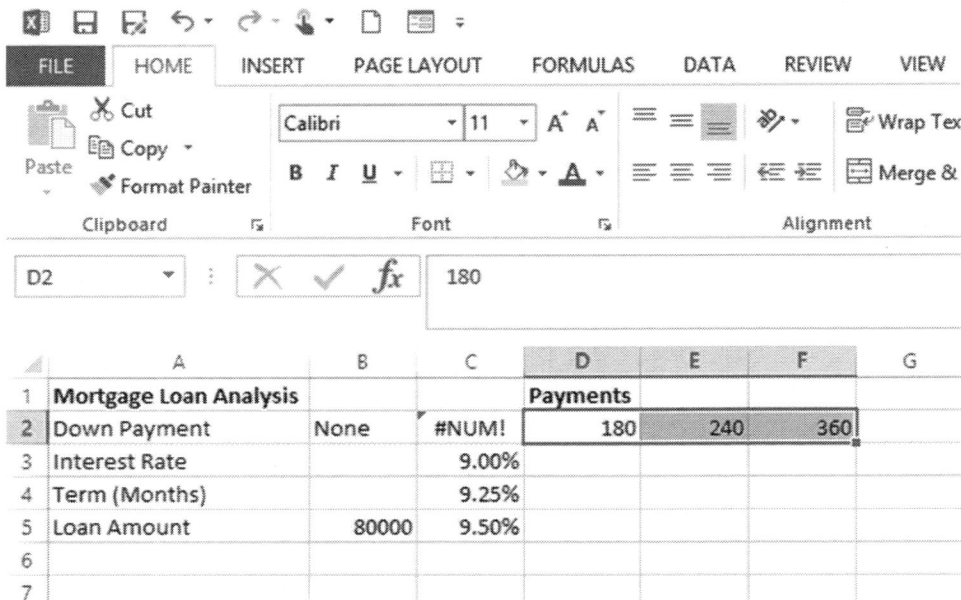

Step 5: Select the range that includes data table values, the formula, and the area where Excel will display the results. In this example, the range is C2:D5.

Step 6: Select the Data tab from the Ribbon.

Step 7: Select What If Analysis.

Step 8: Select Data Table.

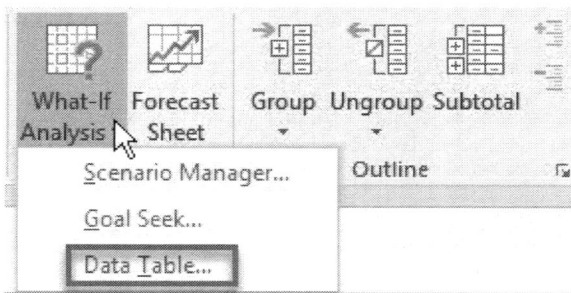

Step 9: Select the Row input cell in the formula. In this example, the cell B4 is the Row Input cell.

Step 10: Select the Column input cell in the formula. In this example, the cell B3 is the Column Input cell.

160

	A	B	C	D	E	F	G	H	I	J
1	Mortgage Loan Analysis			Payments						
2	Down Payment	None	#NUM!	180	240	360				
3	Interest Rate		9.00%							
4	Term (Months)		9.25%							
5	Loan Amount	80000	9.50%							
6										
7										
8										
9										
10										

Data Table ? ×

Row input cell: $B$4

Column input cell: $B$3

OK   Cancel

Step 11: Select OK.

For each possible value for the variable listed in the data table, Excel displays the results.

	A	B	C	D	E	F
1	Mortgage Loan Analysis			Payments		
2	Down Payment	None	#NUM!	180	240	360
3	Interest Rate		9.00%	811.41327	719.7808	643.6981
4	Term (Months)		9.25%	823.35383	732.6935	658.1403
5	Loan Amount	80000	9.50%	835.37975	745.705	672.6834

You may want to format the cells to show the results with the desired formatting (such as currency in this example).

Chapter 19 – Managing Your Data

In this chapter, you will learn how to transpose data from rows to columns. You will also learn about the Text to Columns feature. This chapter explains how to check for duplicates and create data validation rules. You will also learn how to consolidate data.

Transposing Data from Rows to Columns

To transpose data, use the following procedure.

Step 1: Copy the range of cells you want to transpose. This feature will not work if you cut the cells.

Step 2: Place your cursor in the new location and right-click.

Step 3: Select Transpose from the Paste Options on the context menu.

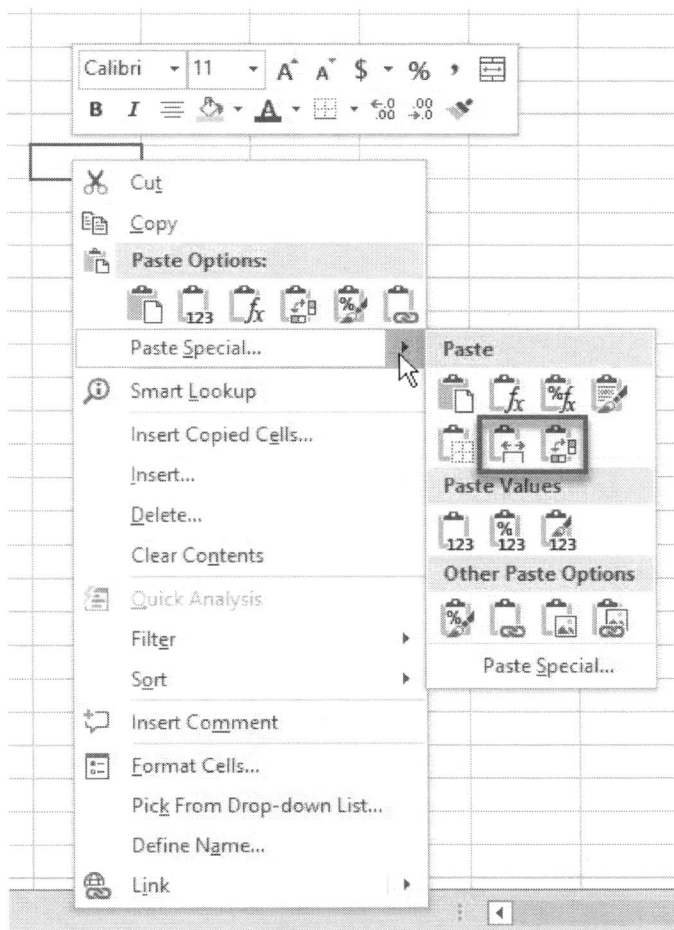

To convert text to columns, use the following procedure.

Step 1: Paste text from another application.

Step 2: Select the text.

Step 3: Select the Data tab from the Ribbon.

Step 4: Select Text to Columns.

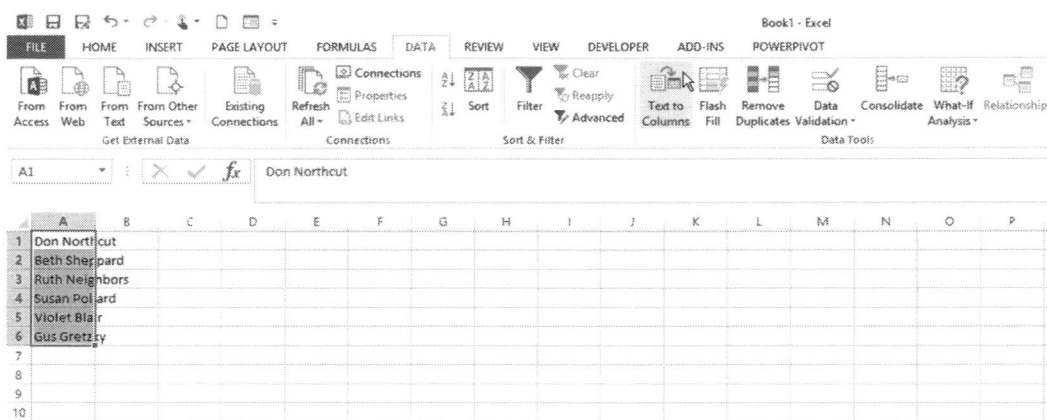

Step 5: In the Convert Text to Columns Wizard, choose the file type that best describes your data. Select Next.

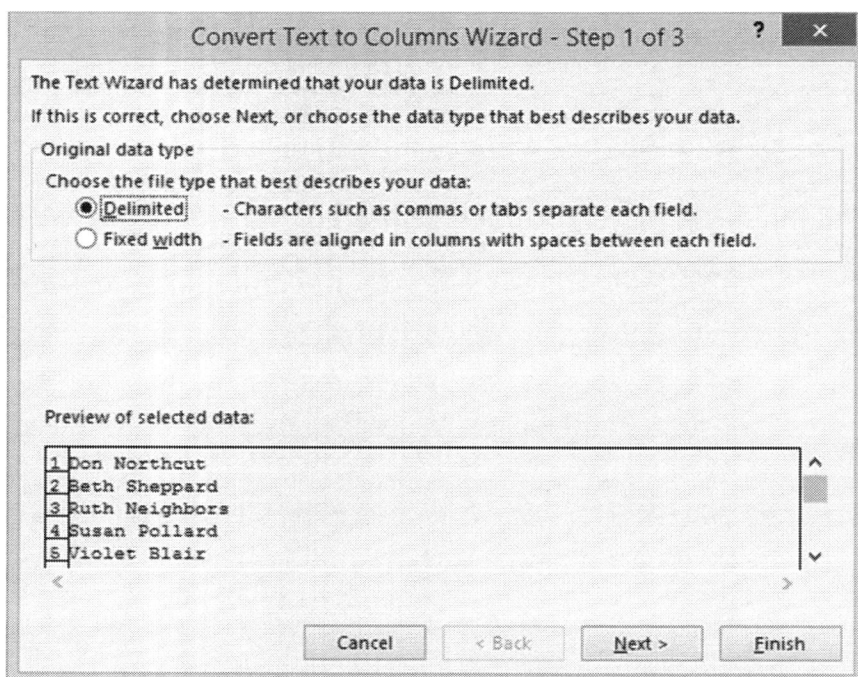

Step 6: In the next screen, select the type of divider. In this example, a space separates the items we want to convert to columns. Your text could be divided by almost any character. Select Next.

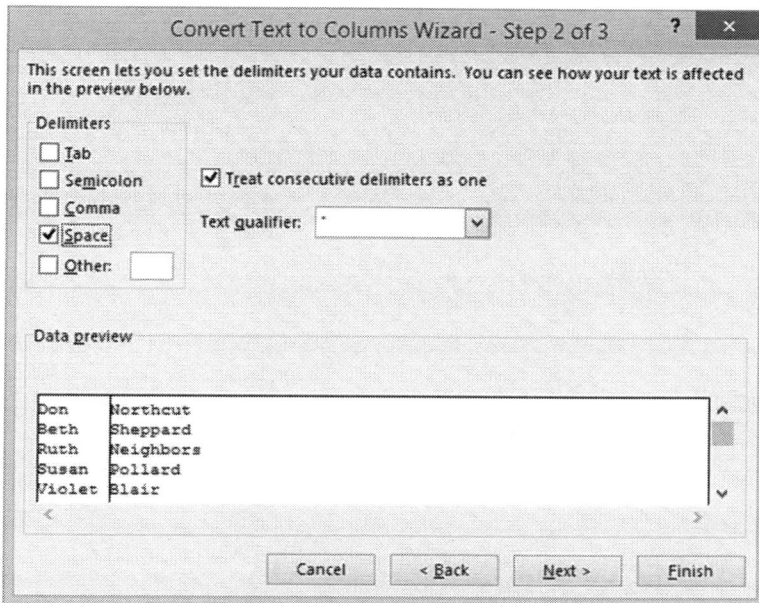

Step 7: In the next screen, you see a preview of the data converted to columns. For each column:

- Define the data format (General, Text, Date) or choose to skip that column.
- Enter or select the destination on the worksheet.

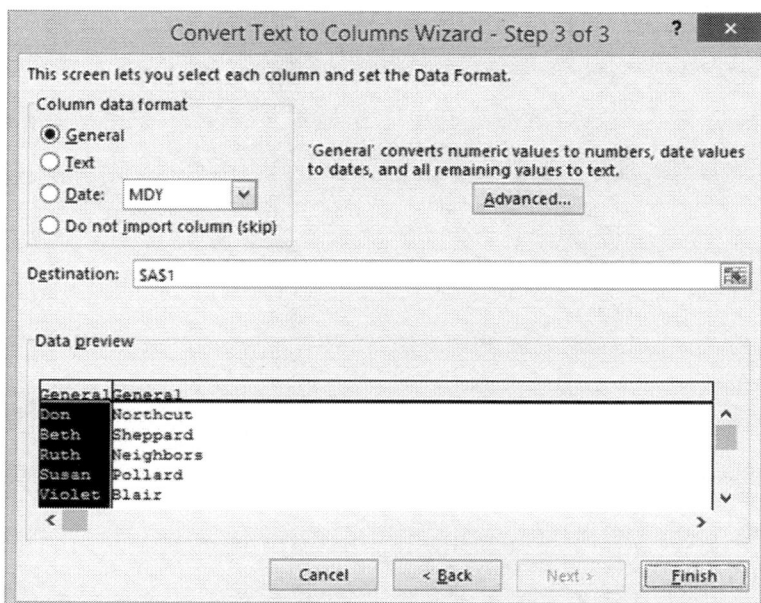

Using the Advanced button, you can also choose your settings for numeric data.

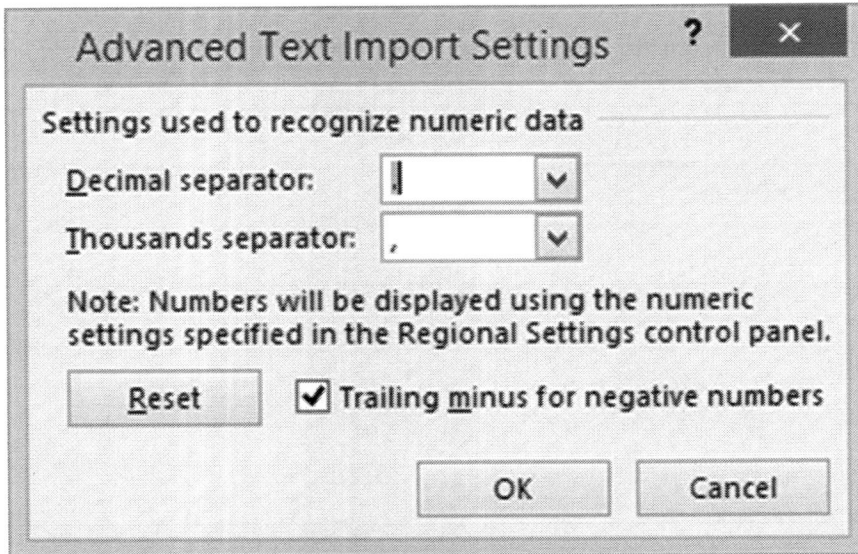

Step 8: When you have finished, select Finish.

You can now work with your data as separate columns.

Checking for Duplicates

To check for duplicate data, use the following procedure.

Step 1: Highlight the area from which you want to remove duplicates.

Step 2: Select the Data tab from the Ribbon.

Step 3: Select Remove Duplicates.

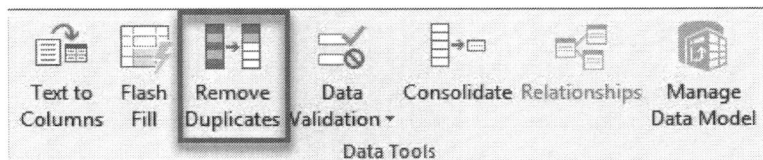

Step 4: Select the columns you want to check for duplicates. The Select All and Unselect All tools can help you manage a large list of columns. The My Data has Headers box indicates whether the list includes header rows.

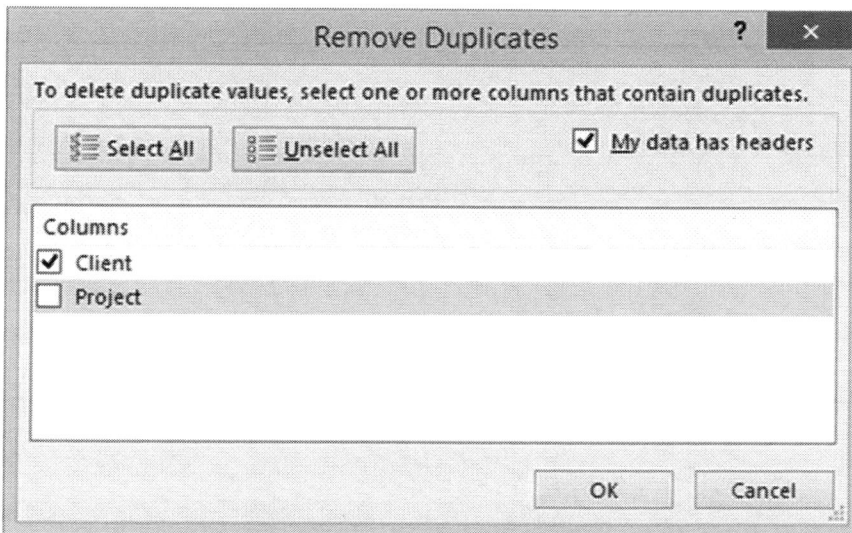

Step 5: Select OK.

Excel notifies you of how many duplicates are removed.

Creating Data Validation Rules

To create a data validation rule, use the following procedure.

Step 1: Select the cells where you want to apply the data validation rule.

Step 2: Select the Data tab from the Ribbon.

Step 3: Select Data Validation.

Step 4: Select Data Validation from the drop-down list.

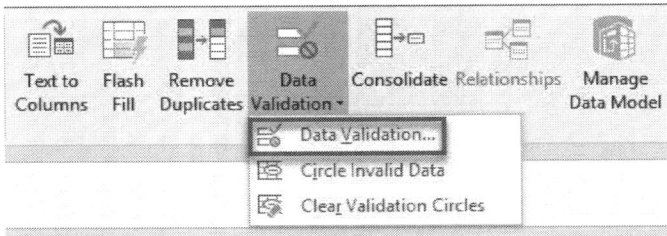

Step 5: On the Settings tab of the Data Validation dialog box, set up the Validation Criteria. Use the drop down lists to help you build your criteria. In this example, we are requiring a three-digit number.

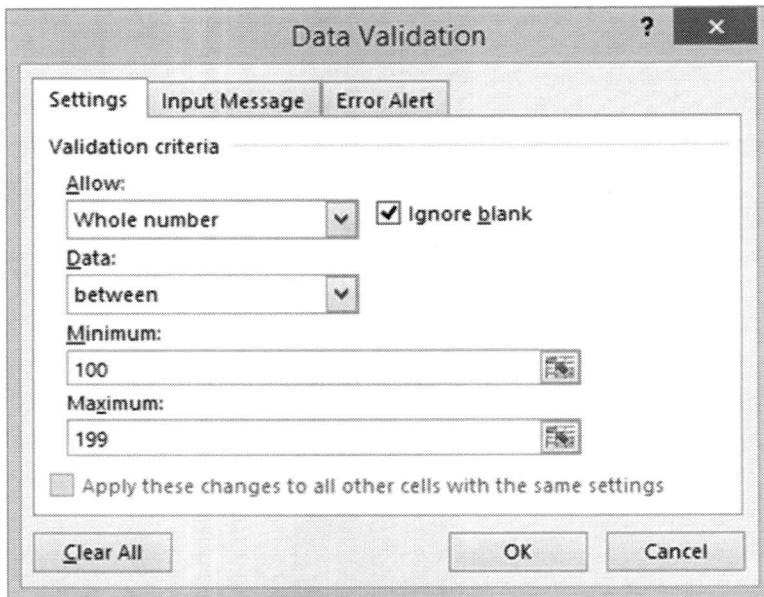

Step 6: Select the Input Message tab.

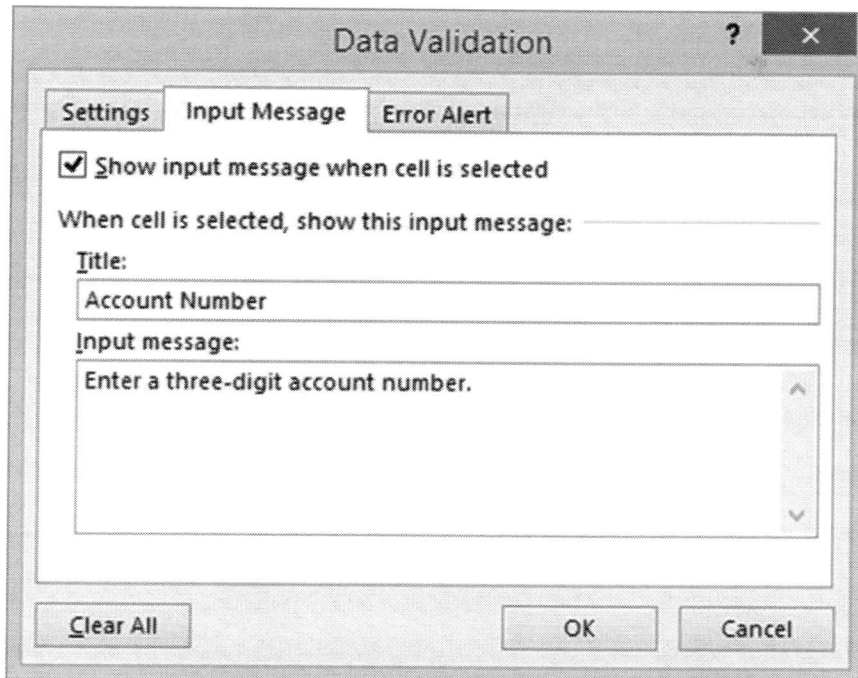

Step 7: Enter a Title and Message that the user will see when he or she selects the cell.

Step 8: Select the Error Alert tab.

Step 9: Select the Style of error from the drop-down list. Enter a Title and Error message to display if the user enters invalid data.

Step 10: Select OK.

Review what happens when you break the validation rule.

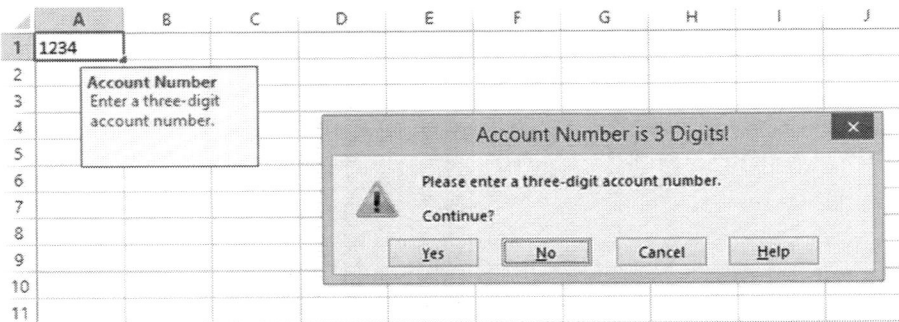

Consolidating Data

To consolidate data, use the following procedure.

Step 1: Select the starting cell where you want to display the consolidated data. Make sure to leave enough room for the consolidated data, so that you do not overwrite other information. In this example, choose the top left cell in Sheet 3.

Step 2: Select the Data tab from the Ribbon.

Step 3: Select Consolidate.

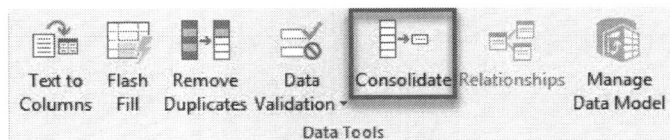

Step 4: In the Consolidate dialog box, do the following:

- Select the Function from the drop-down list. In this example, use Average.

- Select the Reference for each worksheet you are consolidating. If the worksheet is in another workbook, select Browse to open it. Select the cells to include in the consolidation from the first worksheet and select Add. Repeat for each reference.

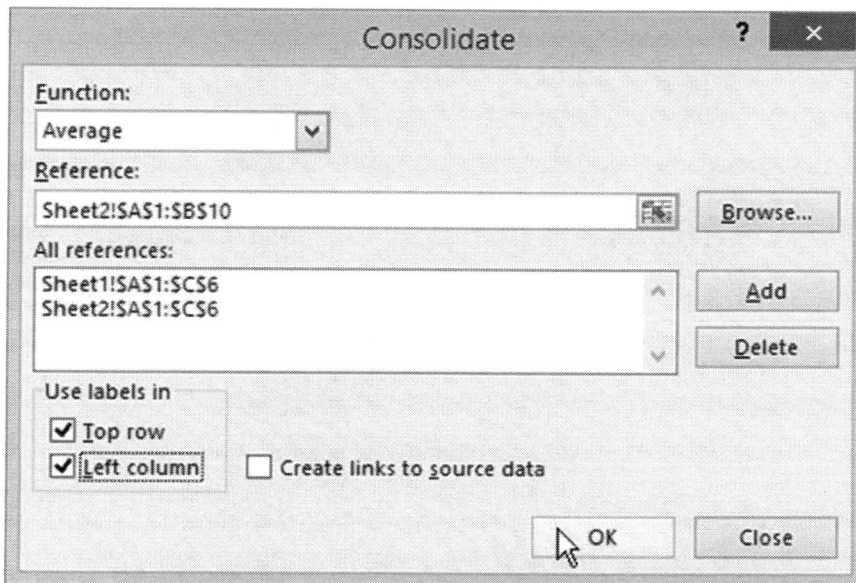

Consolidate ? ✕

Function:
Average ▾

Reference:
Sheet2!$A$1:$B$10   [🔳]   Browse...

All references:
Sheet1!$A$1:$C$6
Sheet2!$A$1:$C$6   Add

     Delete

Use labels in
☑ Top row
☑ Left column   ☐ Create links to source data

OK   Close

Excel has some powerful tools to help you quickly group and outline your data. In this chapter, you will learn how to group your data. You will also learn about adding subtotals to a list of data. This chapter explains outlining data. It also explains how to view grouped and outlined data.

Grouping Data

To create a group, use the following procedure.

Step 1: Select the range of cells you want to group.

Step 2: Select the Data tab from the Ribbon.

Step 3: Select Group.

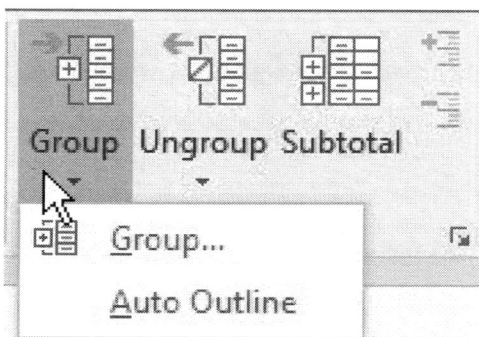

Adding Subtotals

To add subtotals, use the following procedure.

Step 1: Make sure that each column of data has a label in the first row. It must also contain similar facts. Do not include any blank rows or columns.

Step 2: Select the Data tab from the Ribbon.

Step 3: Select Subtotal.

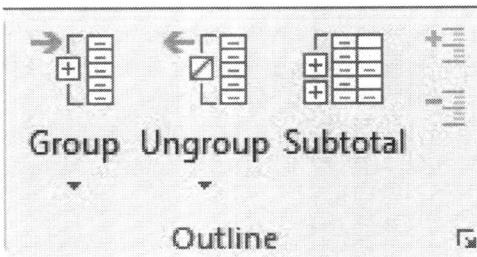

Group Ungroup Subtotal

Outline

Step 4: In the Subtotal dialog box, select the locations for the subtotals from the At each change in drop down list.

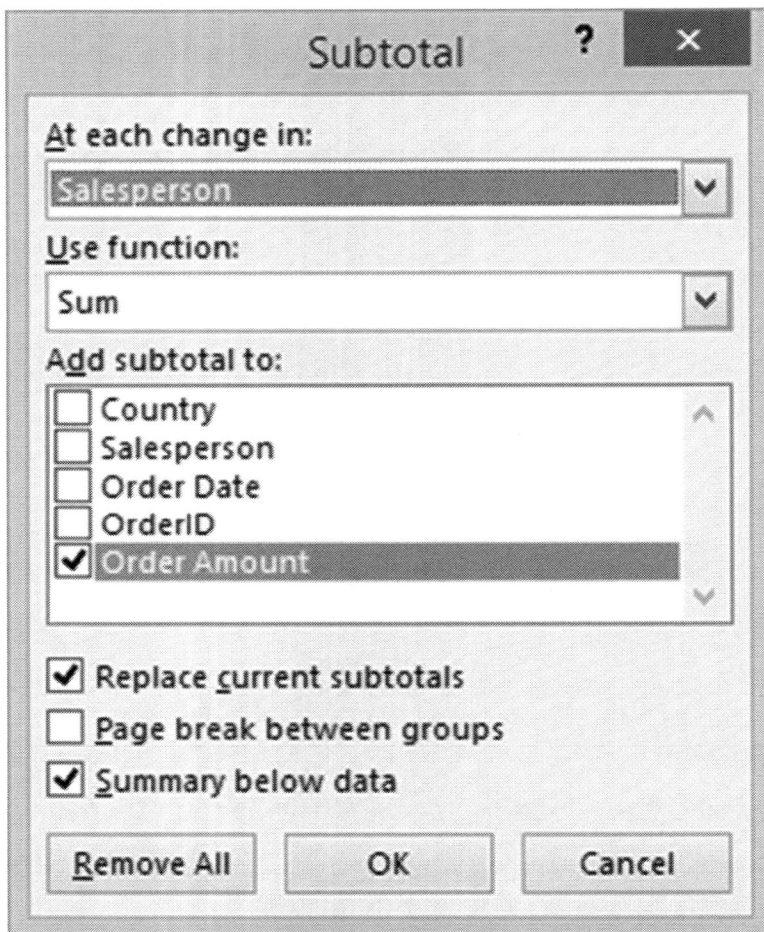

Step 5: Select the function to use in the subtotal fields from the Use Function drop down list.

Step 6: Check the boxes that correspond to your column headers for which column(s) to subtotal.

174

Step 7: Check the boxes to indicate the other formatting options by checking or clearing the Replace current subtotals, Page break between groups, and Summary below data.

Step 8: Select OK.

		A	B	C	D	E	F
34		UK	Buchanan	12/19/2008	10378	$103.20	
35		UK	Buchanan	12/9/2008	10372	$9,210.90	
36		UK	Buchanan	11/27/2008	10358	$429.40	
37		UK	Buchanan	11/26/2008	10359	$3,471.68	
38		UK	Buchanan	10/25/2008	10333	$877.20	
39		UK	Buchanan	10/18/2008	10320	$516.00	
40		UK	Buchanan	9/10/2008	10297	$1,420.00	
41		UK	Buchanan	8/9/2008	10269	$642.20	
42		UK	Buchanan	7/23/2008	10254	$556.62	
43		UK	Buchanan	7/16/2008	10248	$440.00	
44			**Buchanan Total**			$68,792.25	
45		USA	Callahan	5/1/2010	11056	$3,740.00	
46		USA	Callahan	4/27/2010	11034	$539.40	
47		USA	Callahan	4/24/2010	11046	$1,485.80	
48		USA	Callahan	4/22/2010	11036	$1,692.00	
49		USA	Callahan	4/21/2010	10986	$2,220.00	
50		USA	Callahan	4/17/2010	10998	$686.00	
51		USA	Callahan	4/13/2010	10997	$1,885.00	
52		USA	Callahan	4/13/2010	11007	$2,633.90	
53		USA	Callahan	4/10/2010	10977	$2,233.00	

Outlining Data

To create an outline, use the following procedure.

Step 1: Select the range of cells to include in the outline.

Step 2: Select the Data tab from the Ribbon.

Step 3: Select the small square in the corner of the Outline group.

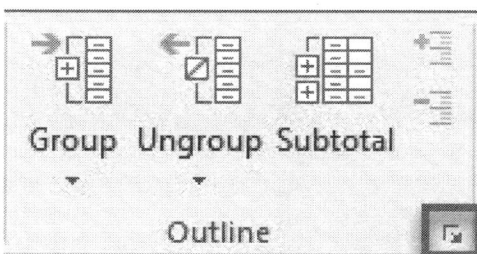

Step 4: In the Settings dialog box, check the direction of the summary rows and columns.

175

Step 5: Check the Automatic styles box to have Excel automatically apply styles to the outline.

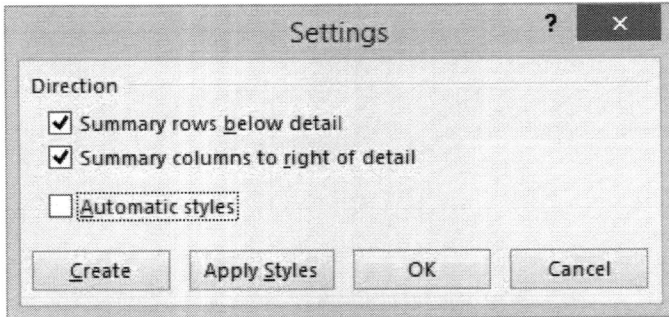

Step 6: Select Create.

Viewing Grouped and Outlined Data

To work with grouped or outlined data.

The Hide Detail icon allows you to quickly hide the detail data.

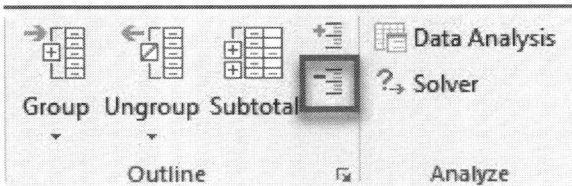

The + icons indicate hiding detail data.

The Show Detail icon allows you to quickly show the detail data. You can select the Show Detail icon multiple times to continue expanding the current level.

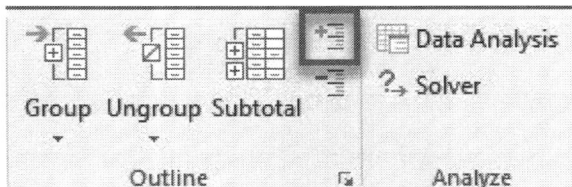

The minus icons allow you to collapse individual groups. The plus icons allow you to expand individual groups.

The numbers in the top left corner indicate a level. Click on a number to show that level.

	Country	Salesperson	Order Date	OrderID	Order Amount
44		0 Buchanan Total			
144		0 Callahan Total			
262		0 Davolio Total			
304		0 Dodsworth Total			
397		0 Fuller Total			
465		0 King Total			
591		0 Leverling Total			
743		0 Peacock Total			
809		0 Suyama Total			
810		0 Grand Total			
811					

This chapter introduces you to the Information tab on the Backstage View. You will learn about marking a workbook as final, which makes the workbook read-only. You will also learn about permissions – both encrypting the workbook with a password and restricting permissions. This chapter explains how to protect both the current sheet and an entire workbook's structure. You will also learn how to add a digital signature, which is helpful if your workbook contains macros that you want to share with others. Then we will move on to exploring the Excel options dialog box, where you can set advanced options and properties. We will look at managing versions, which can help you recover unsaved work if you have Auto Save turned on. Finally, we will look at saving your workbook as a template to simplify new workbook creation.

Marking a Workbook as Final

To mark a workbook as final, use the following procedure.

Step 1: Select the File tab from the Ribbon to open the Backstage View.

Step 2: Select Protect Workbook.

Step 3: Select Mark as Final.

Step 4: Excel displays a warning message. Select OK to continue.

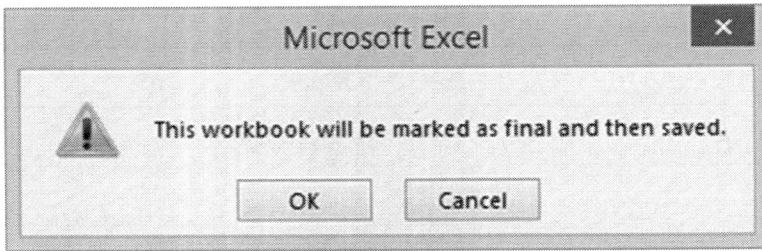

Step 5: Excel displays an information message. Select OK to continue.

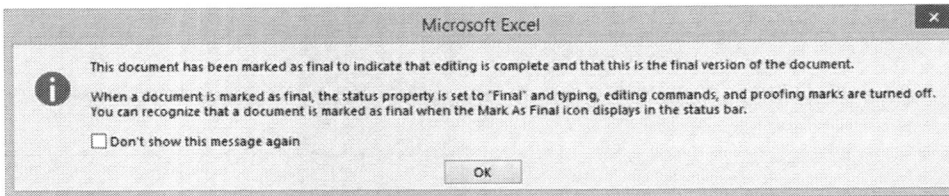

Notice the yellow bar at the top of the workbook to indicate that the workbook has been marked as final.

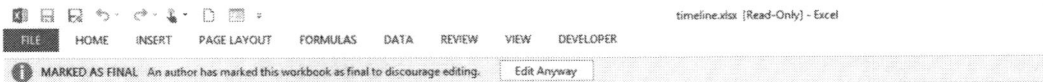

Notice that on the Info tab on the Backstage View, the Permissions area has changed.

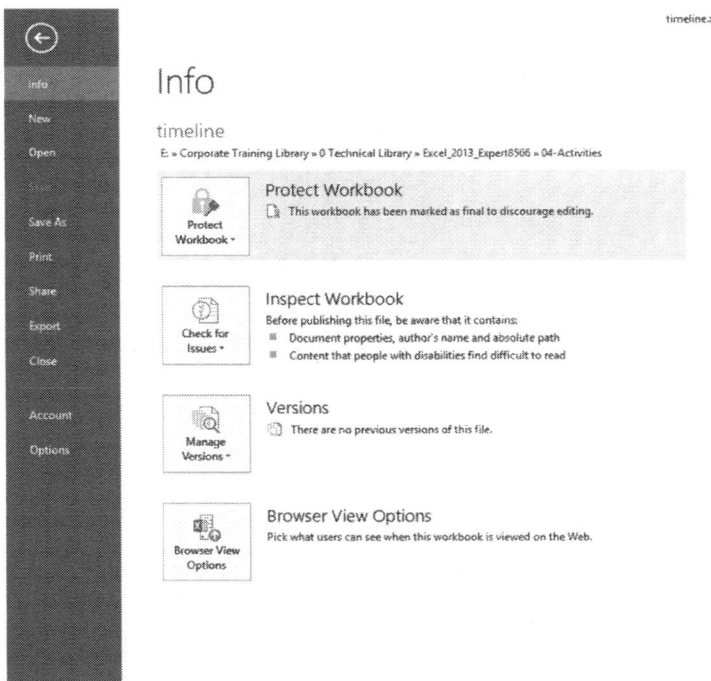

Encrypting with a Password

To encrypt a workbook with a password, use the following procedure.

Step 1: Select the File tab from the Ribbon to open the Backstage View.

Step 2: Select Protect Workbook

Step 3: Select Encrypt with Password.

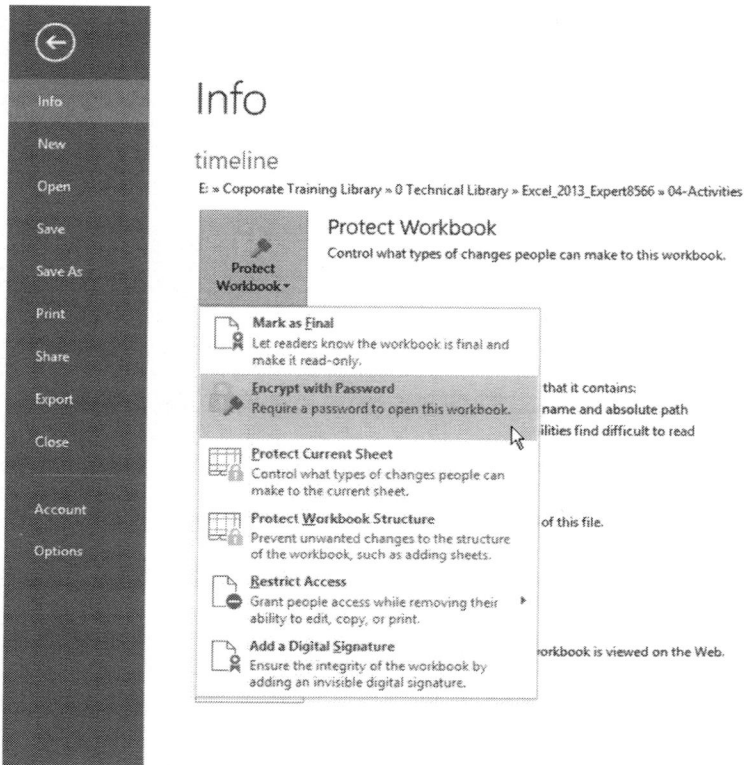

Step 4: In the Encrypt Document dialog box, enter the password that you want to use.

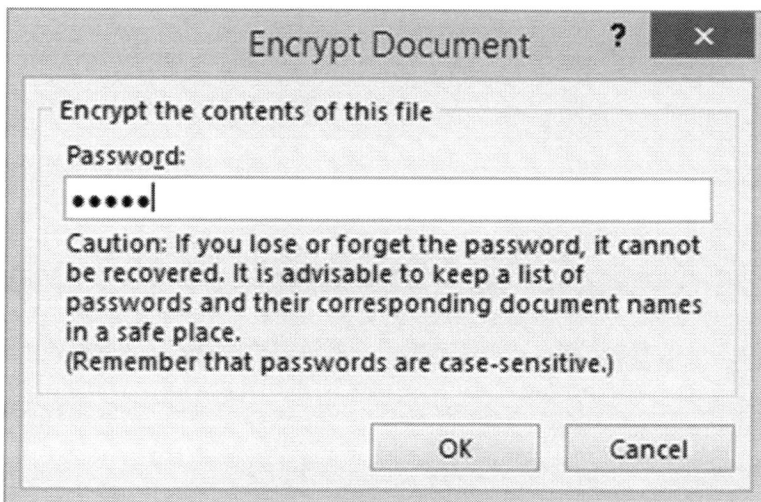

Step 5: In the Confirm Password dialog box, re-enter the password that you want to use to confirm it.

Note that if you want to remove the password protection, you will repeat the process. However, leave the password field blank.

Protecting the Current Sheet or the Workbook Structure

To protect a current sheet of a workbook, use the following procedure.

Step 1: Select the File tab from the Ribbon to open the Backstage View.

Step 2: Select Protect Workbook.

Step 3: Select Protect Current Sheet.

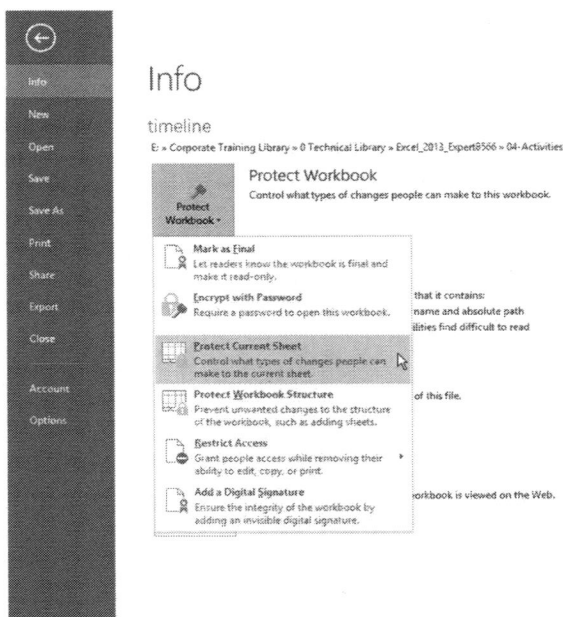

Step 4: Excel displays the Protect Sheet dialog box.

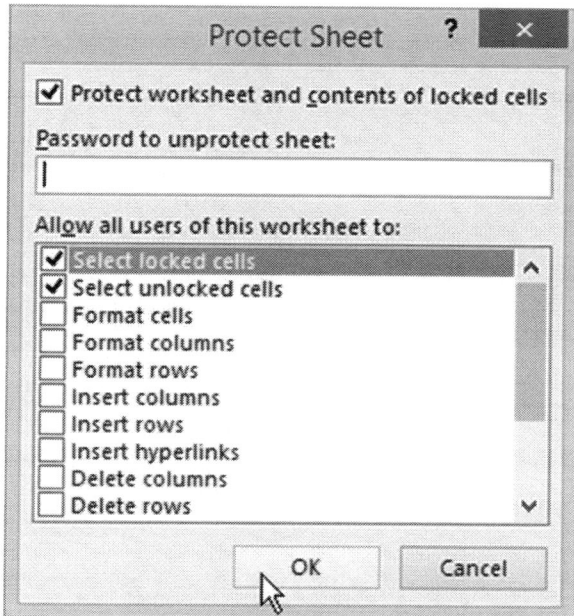

Step 5: You can enter a password if desired to unprotect the sheet.

Step 6: Check the boxes for the actions that you want to allow other users to perform on the sheet.

Step 7: Select OK.

To protect a workbook structure, use the following procedure.

Step 1: Select the File tab from the Ribbon to open the Backstage View.

Step 2: Select Protect Workbook.

Step 3: Select Protect Workbook Structure.

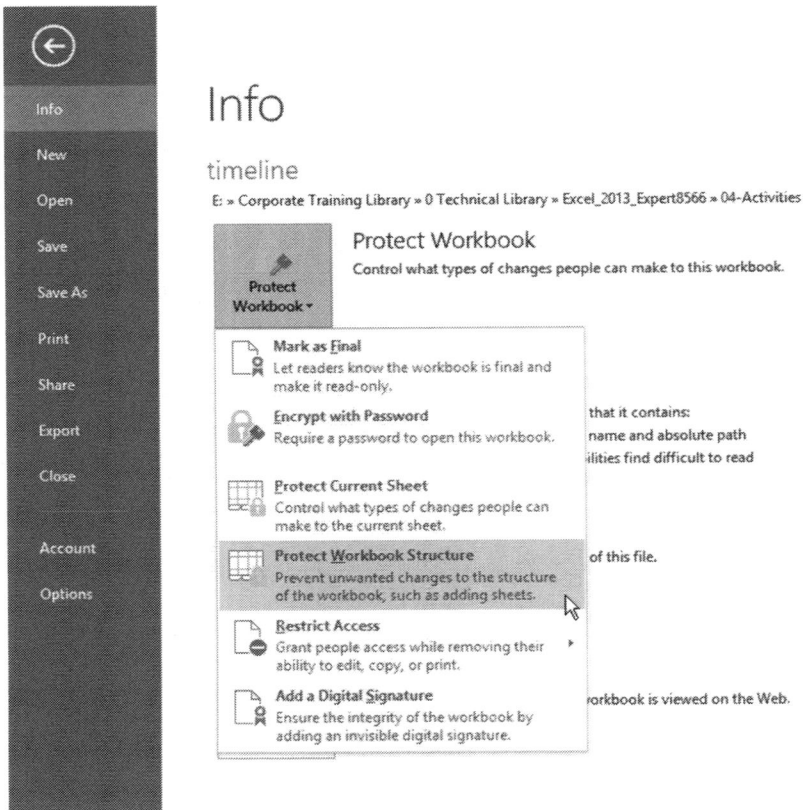

Step 4: Excel displays the Protect Structure and Windows dialog box.

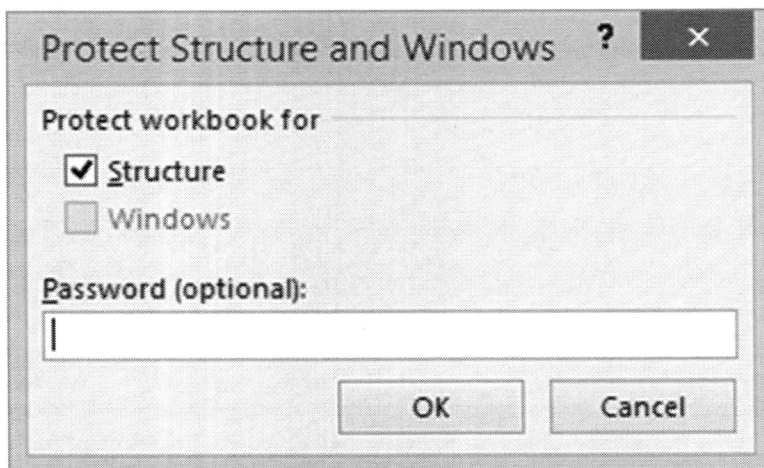

Step 5: Check the boxes for the options you want to protect.

Step 6: You can enter a password if desired to unprotect the workbook.

Step 7: Select OK.

To add a digital signature to a workbook, use the following procedure.

Step 1: Select the File tab from the Ribbon to open the Backstage View.

Step 2: Select Protect Workbook.

Step 3: Select Add a Digital Signature.

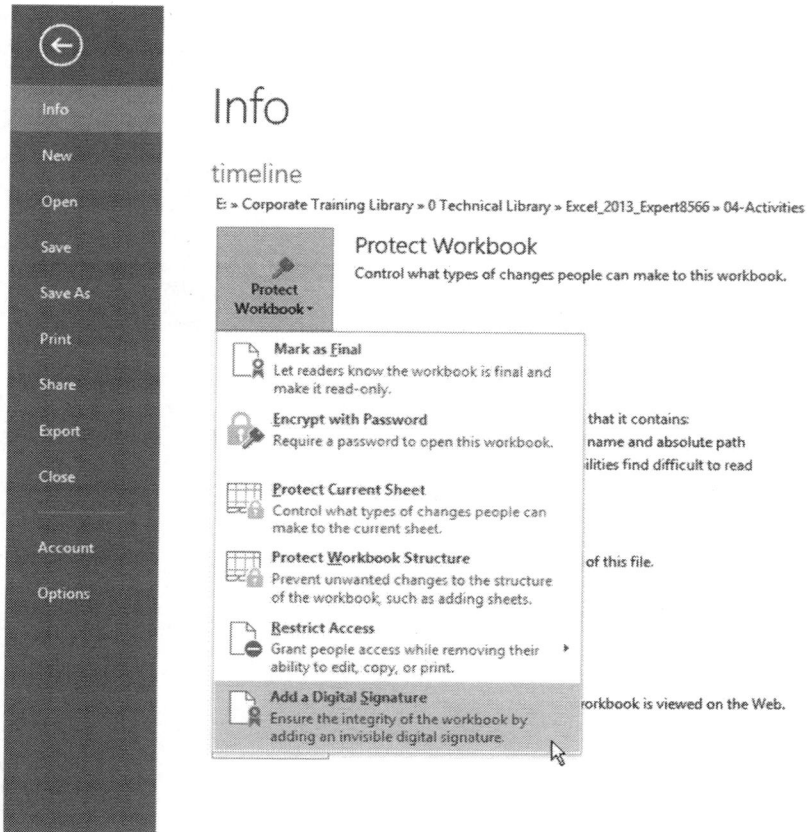

Step 4: Excel may display an informational message. Select OK.

Step 5: In the Sign dialog box, select the Commitment Type from the drop-down list.

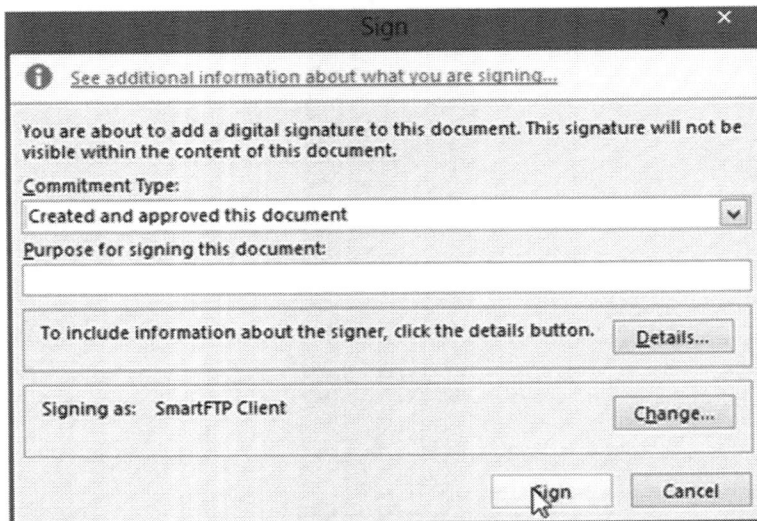

Step 6: Enter a Purpose for signing the document.

Step 7: If you would like to include additional information about the signer, select Details.

Step 8: In the Additional Signing Information dialog box, enter the signature information and select OK.

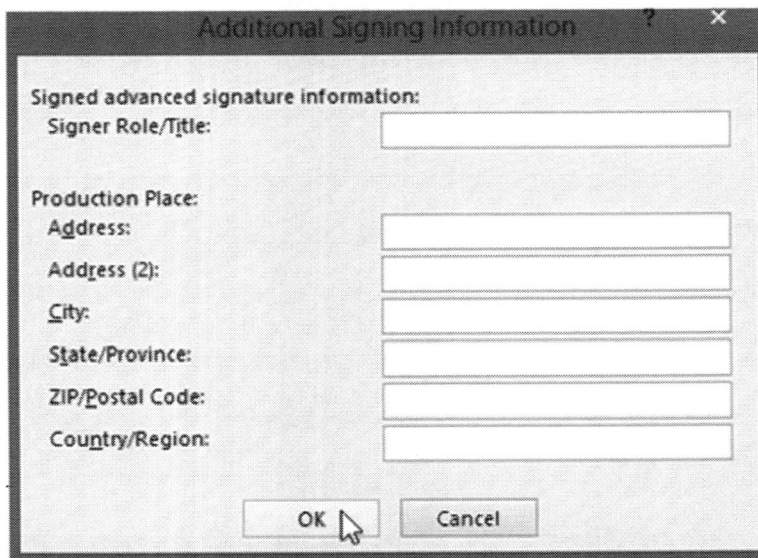

Step 9: Your Signature Certificate should appear in the Signing as area. If not, select Change and choose a new one from the Windows Security dialog box.

Step 10: Select Sign.

Step 11: The Signature Confirmation dialog box displays. Select OK.

Setting Excel Options

To review the options for customizing Excel, use the following procedure.

Step 1: Select the File tab from the Ribbon to open the Backstage view.

Step 2: Select the Options tab on the left.

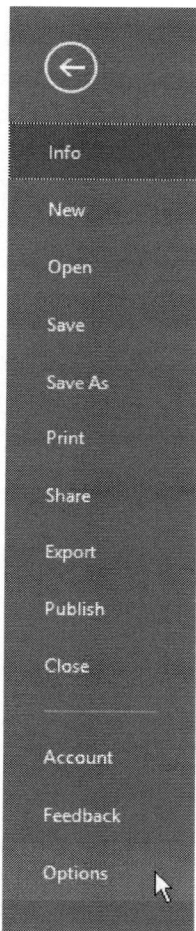

Step 3: Here is the General tab in the Excel Options dialog box. The General tab allows you to change the user interface options. You can enter your name or initials to personalize your copy of Excel.

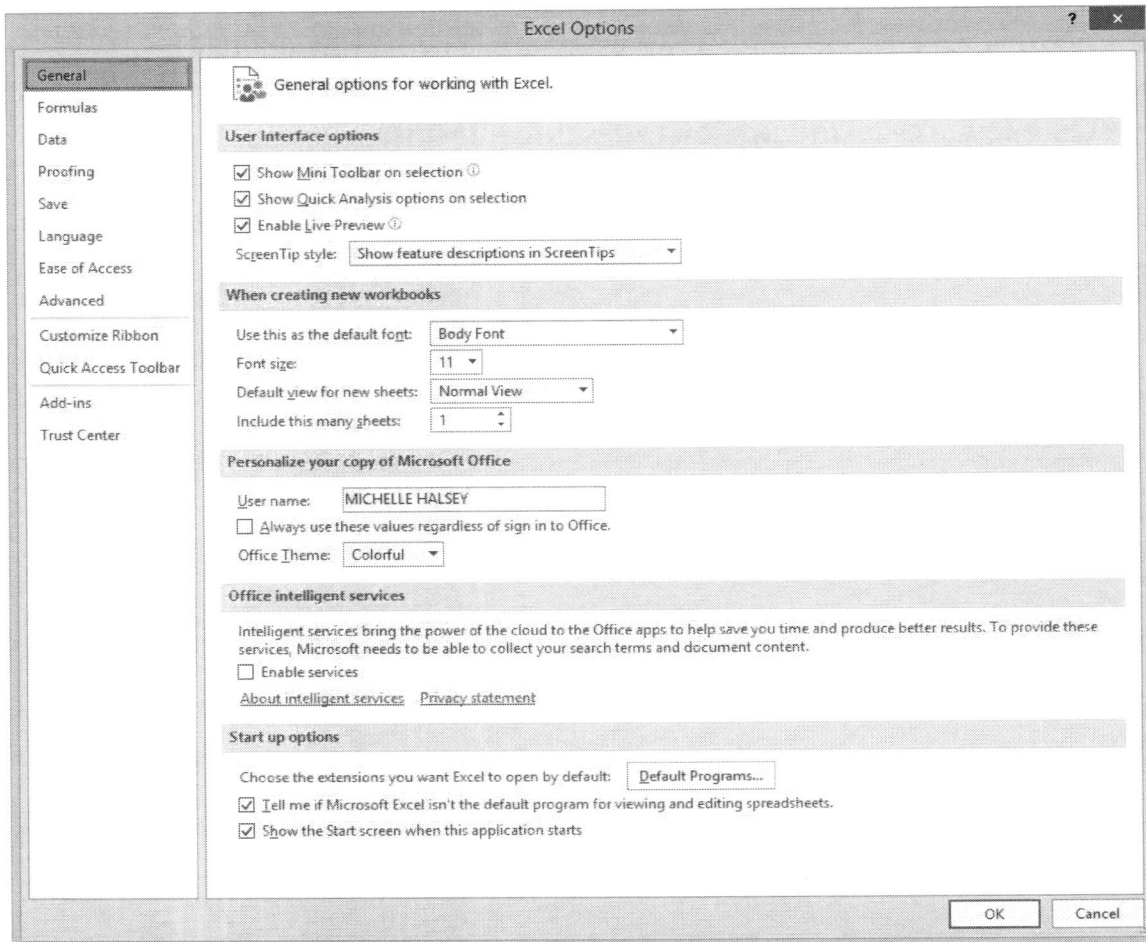

Here is the Formulas tab in the Excel Options dialog box. The Formulas tab controls your Calculation options, how you work with formulas, your error checking options, and the error checking rules.

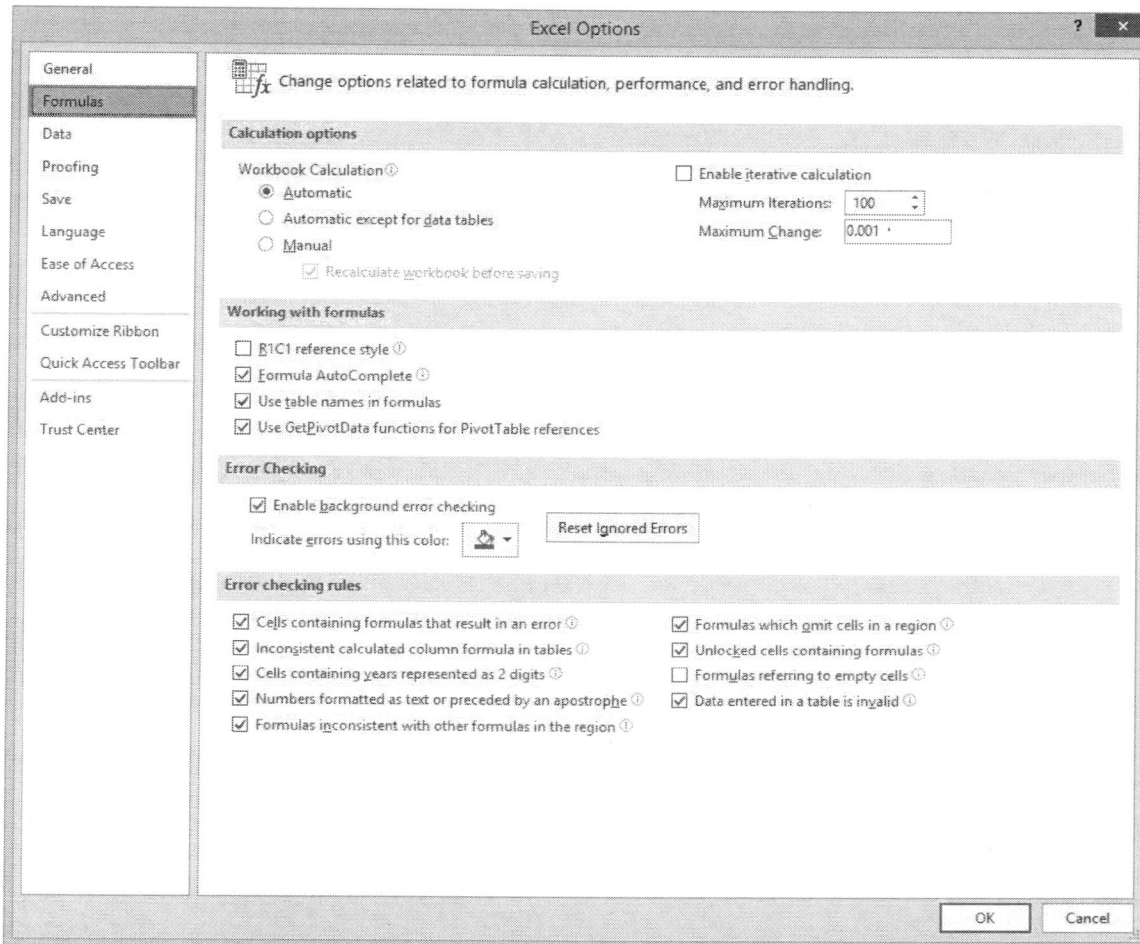

Here is the Data tab in the Excel options dialog box. The Data tab controls your data options and handles the legacy data import wizards.

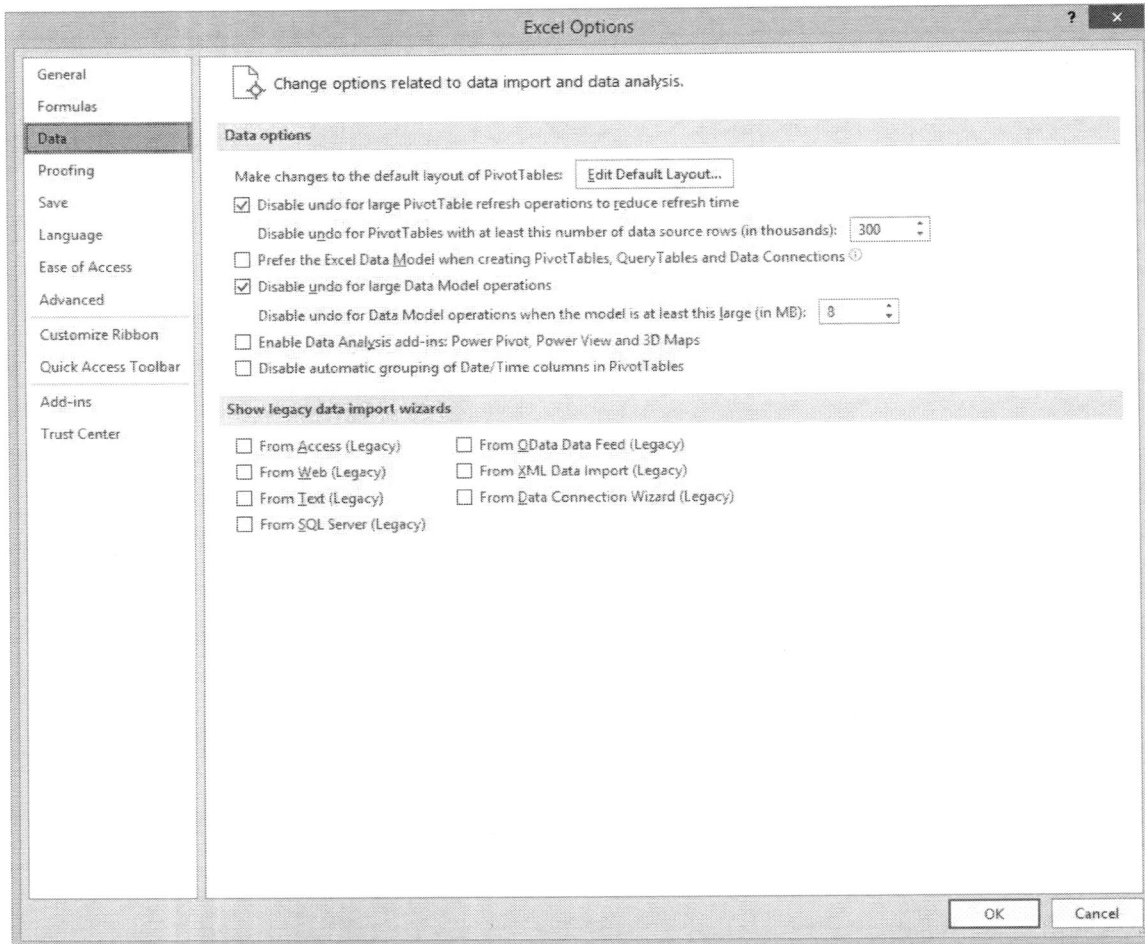

Here is the Proofing tab in the Excel Options dialog box. The Proofing tab allows you to control how Autocorrect works for spelling.

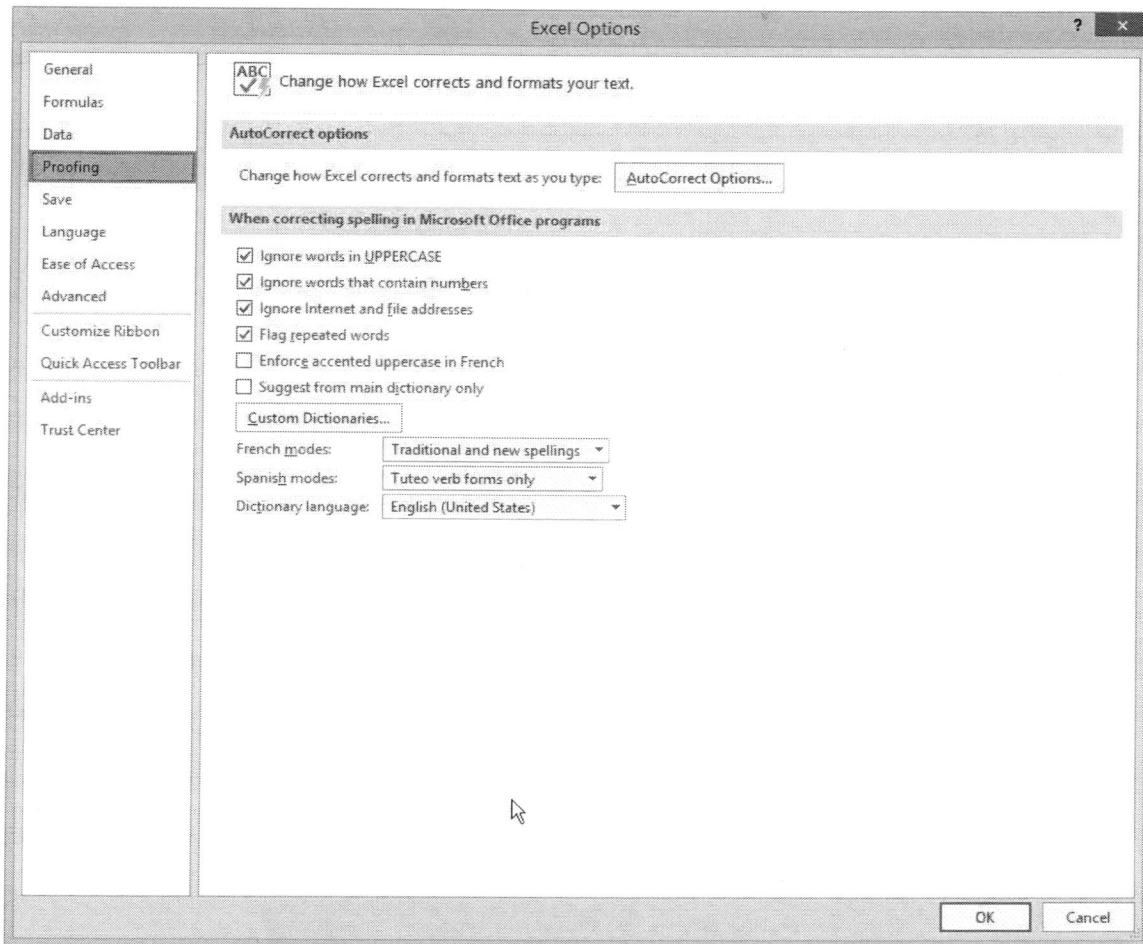

Here is the Save tab in the Excel Options dialog box. The Save tab allows you to control how workbooks are saved.

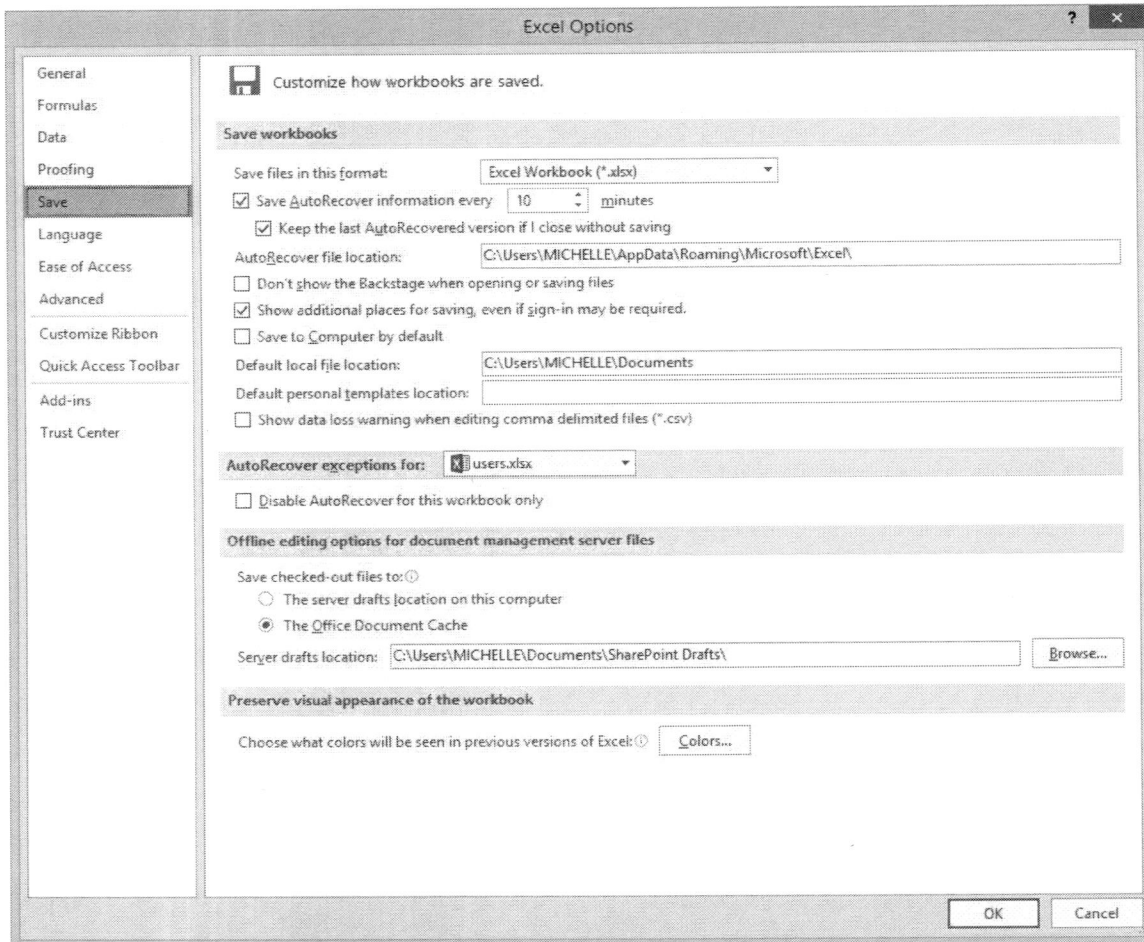

Here is the Language tab in the Excel Options dialog box. The Language tab controls your editing language and your display and help language.

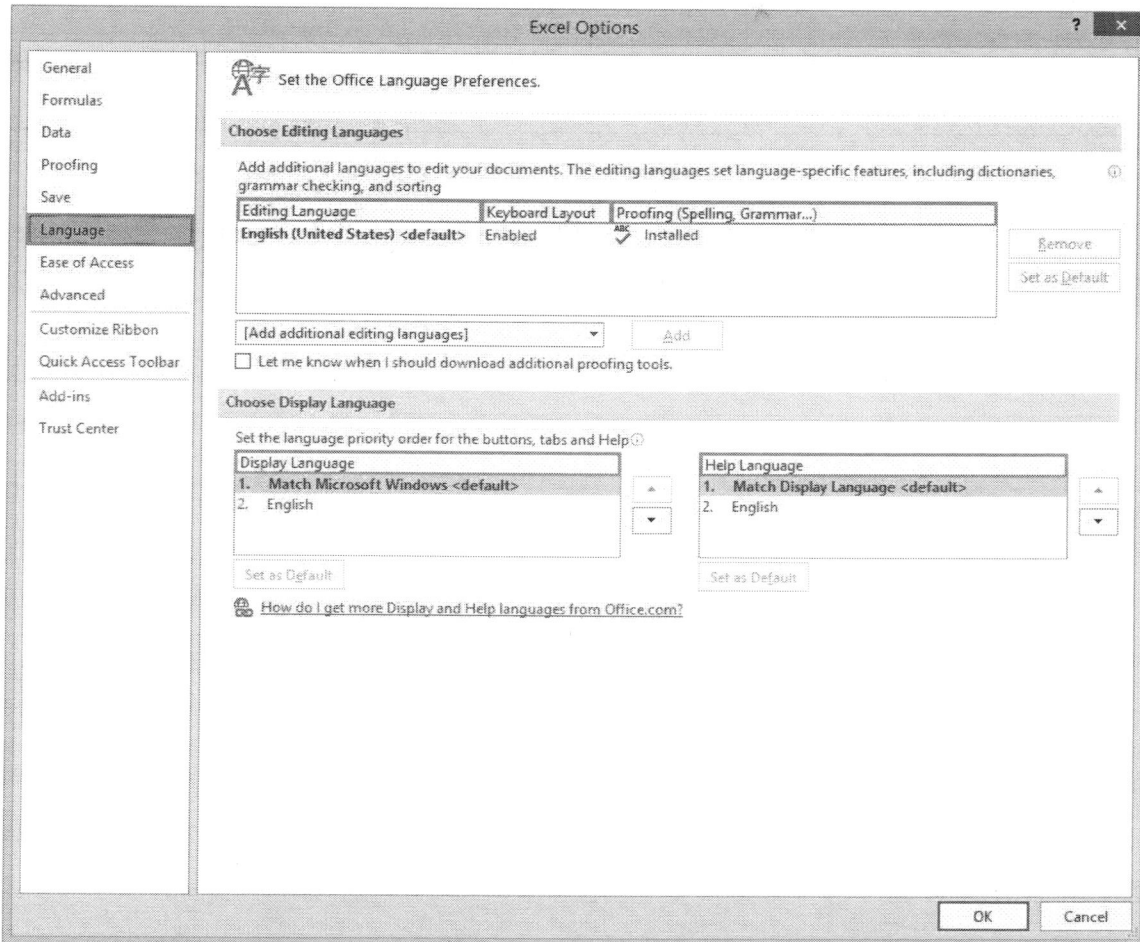

Here is the Ease of Access tab in the Excel Options dialog box. The "Ease of Access" controls feedback options, application display options, and document display options,

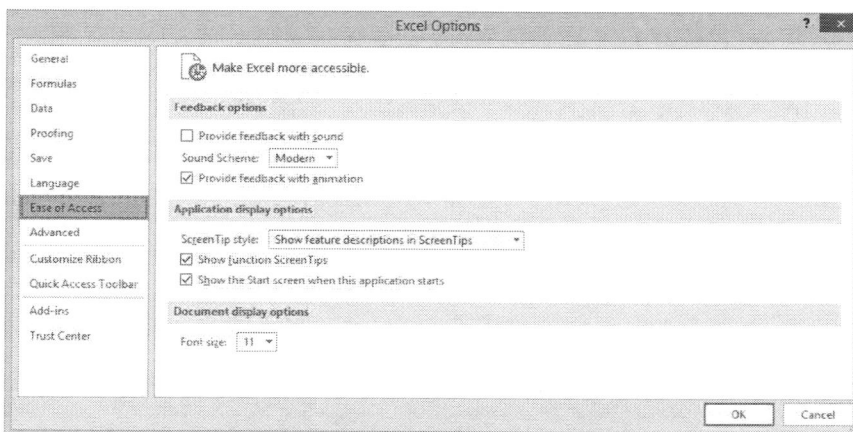

Here is the Advanced tab in the Excel Options dialog box. In the Advanced tab, you can change many editing options, including the default paste option and options when calculating the workbook.

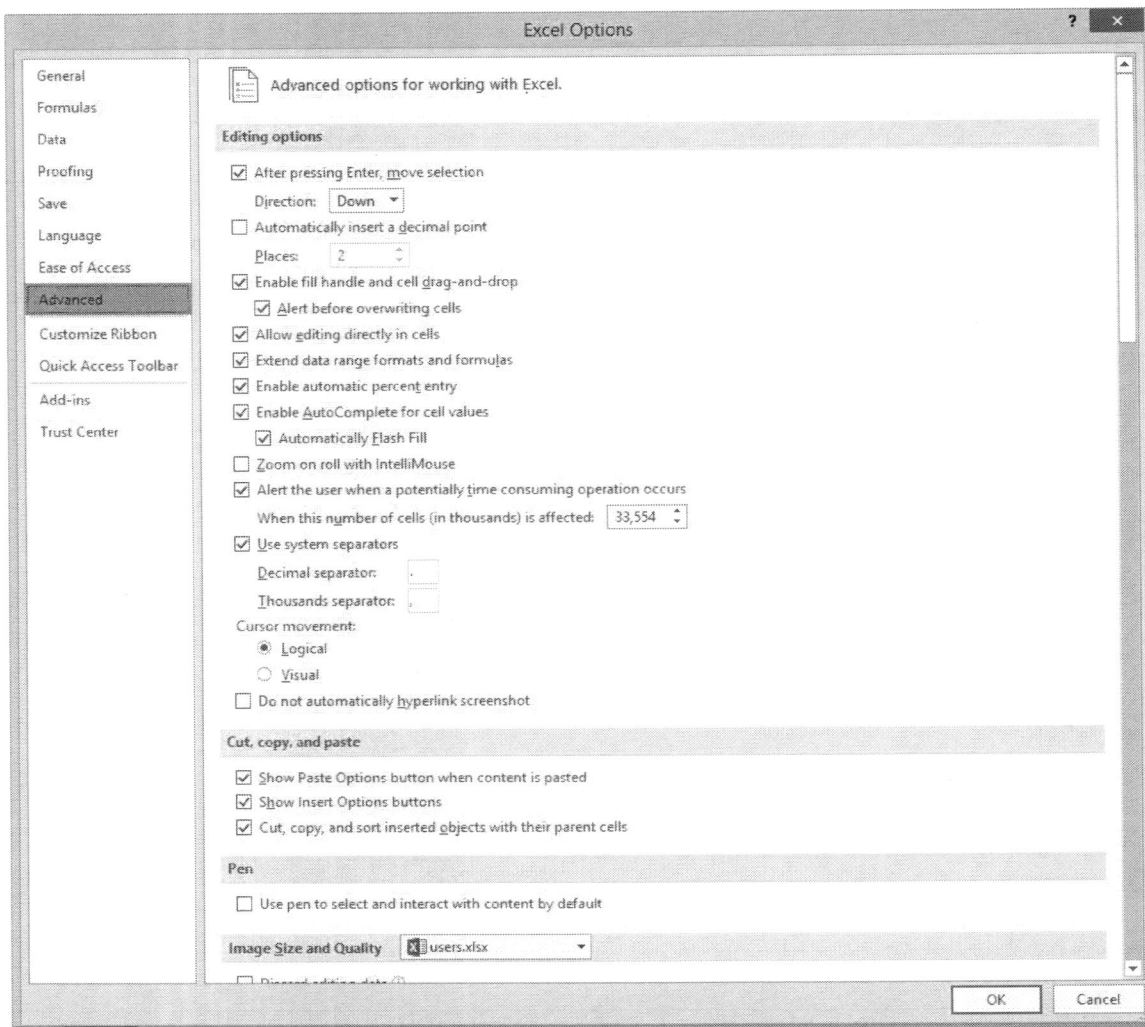

Managing Versions

To use the Manage Versions feature, use the following procedure.

Step 1: Select the File tab from the Ribbon to open the Backstage view.

Step 2: Select Info, if it is not already selected.

Step 3: The Versions area includes the most recent versions of the workbook. You can select one to return to it.

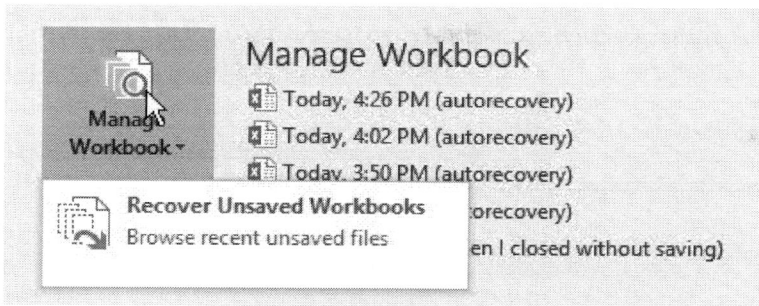

Step 4: Or, select Manage Versions.

Step 5: Select Recover Unsaved Workbooks.

Step 6: The Open dialog box displays a list of your unsaved files. Highlight the file and select Open.

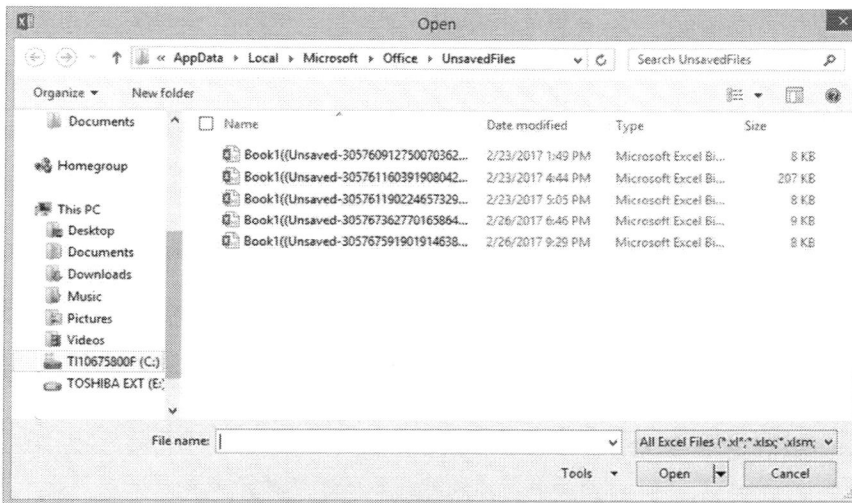

Step 7: Make sure you save the file.

Saving a Workbook as an Excel Template

To set the default template location, use the following procedure.

Step 1: Select the File tab from the Ribbon to open the Backstage view.

Step 2: Select the Options tab on the left.

Step 3: Select the Save tab.

Step 4: In the Default personal templates location field, enter the path to the templates folder you have created.

To save the current workbook as a template, use the following procedure.

Step 1: Select the File tab from the Ribbon to open the Backstage View.

Step 2: Select Save As.

Step 3: In the Save As dialog box, select Excel Template (*xltx) from the Save as Type drop down list.

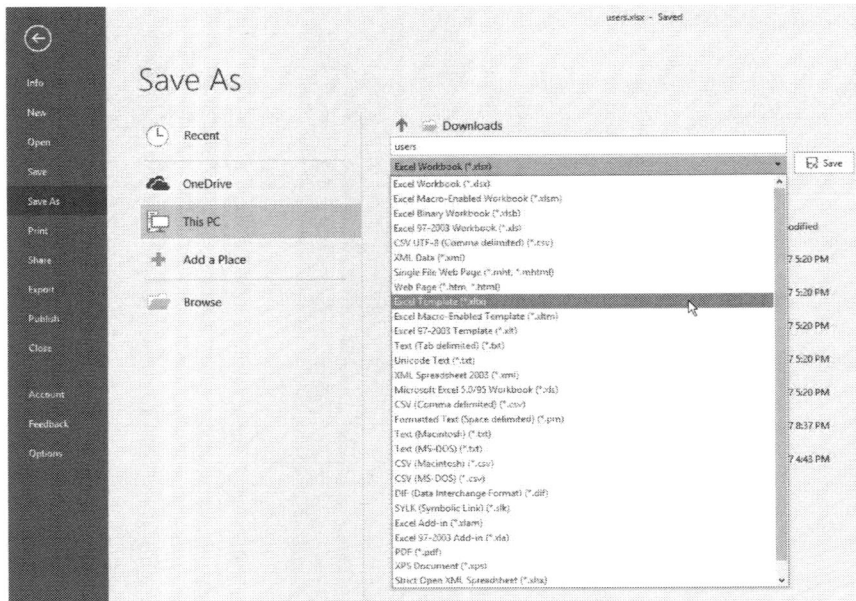

Notice that when you make that selection, the navigation changes to the default templates folder you just set. This is where you will need to save it to make it available for your use when creating new workbooks.

Step 4: Give the template a new name if desired.

Step 5: Select Save.

To create a new file based on the template, use the following procedure.

Step 1: Select the File tab to open the Backstage view.

Step 2: Select New

Step 3: Select Personal.

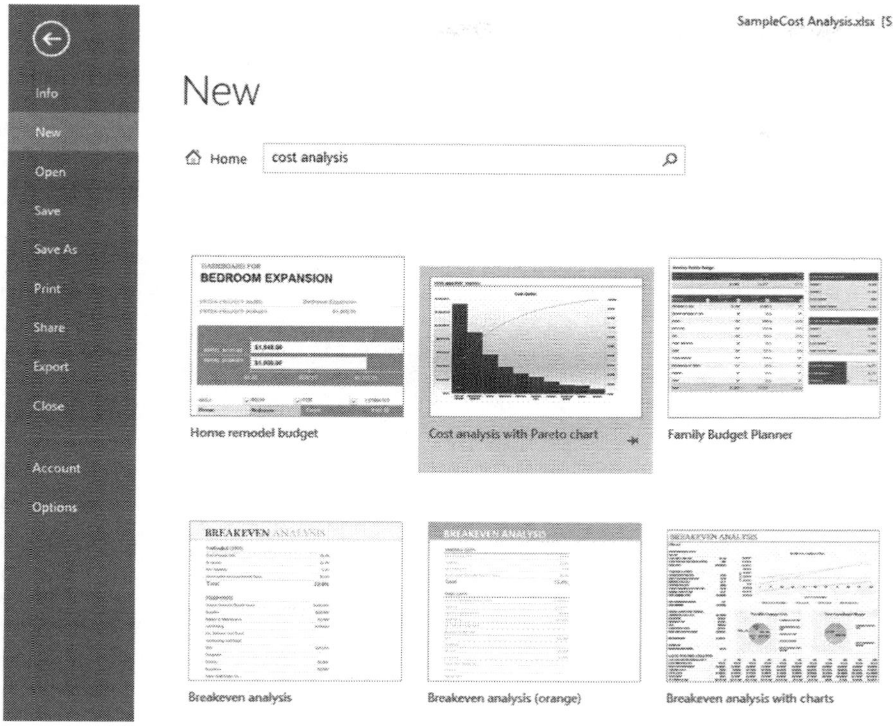

SampleCost Analysis.xlsx [S
Info
New
Open
Save
Save As
Print
Share
Export
Close
Account
Options
New
Home
cost analysis
BEDROOM EXPANSION
Home remodel budget
Cost analysis with Pareto chart
Family Budget Planner
BREAKEVEN ANALYSIS
Breakeven analysis
Breakeven analysis (orange)
Breakeven analysis with charts

Step 4: Select the template you want to use.

This chapter helps you understand the issues concerned with sharing a workbook. First, we will look at how to inspect the workbook for issues. Then, you will learn how to share a workbook and edit a shared workbook. You will also learn about tracking changes to document other users changes and comments. Finally, you will learn how to merge copies of a shared workbook to consolidate the changes.

Inspecting a Document

To inspect a document, use the following procedure.

Step 1: Select the File tab from the Ribbon to open the Backstage View.

Step 2: Select Check for Issues.

Step 3: Select Inspect Document.

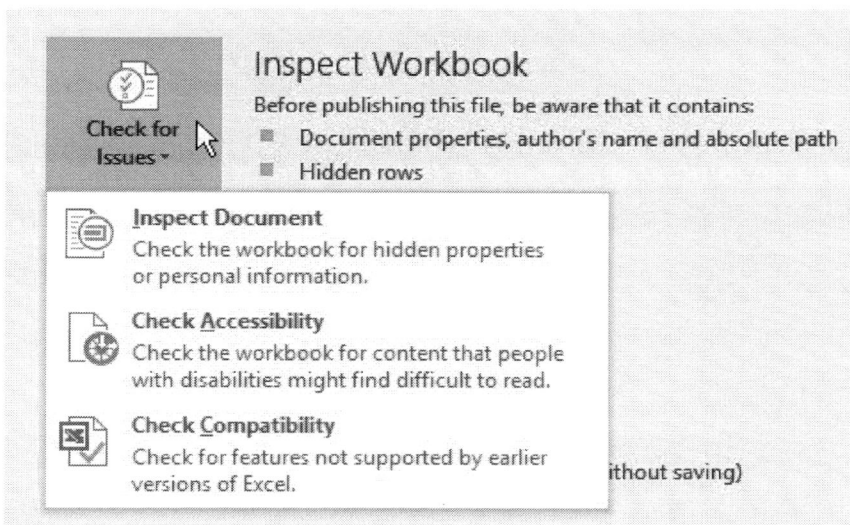

Step 4: In the Document Inspector dialog box, check the boxes for the content you want the Inspector to find. Select Inspect.

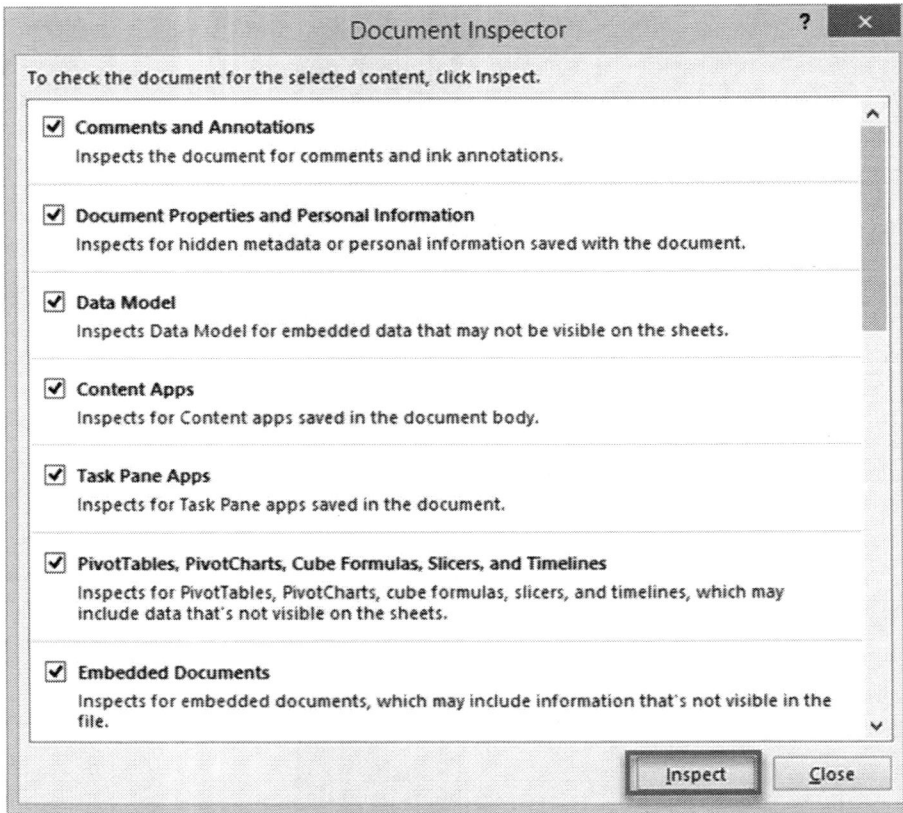

Document Inspector

To check the document for the selected content, click Inspect.

☑ **Comments and Annotations**
　　Inspects the document for comments and ink annotations.

☑ **Document Properties and Personal Information**
　　Inspects for hidden metadata or personal information saved with the document.

☑ **Data Model**
　　Inspects Data Model for embedded data that may not be visible on the sheets.

☑ **Content Apps**
　　Inspects for Content apps saved in the document body.

☑ **Task Pane Apps**
　　Inspects for Task Pane apps saved in the document.

☑ **PivotTables, PivotCharts, Cube Formulas, Slicers, and Timelines**
　　Inspects for PivotTables, PivotCharts, cube formulas, slicers, and timelines, which may include data that's not visible on the sheets.

☑ **Embedded Documents**
　　Inspects for embedded documents, which may include information that's not visible in the file.

[Inspect]　[Close]

Step 5: The Document Inspector displays the inspection results.

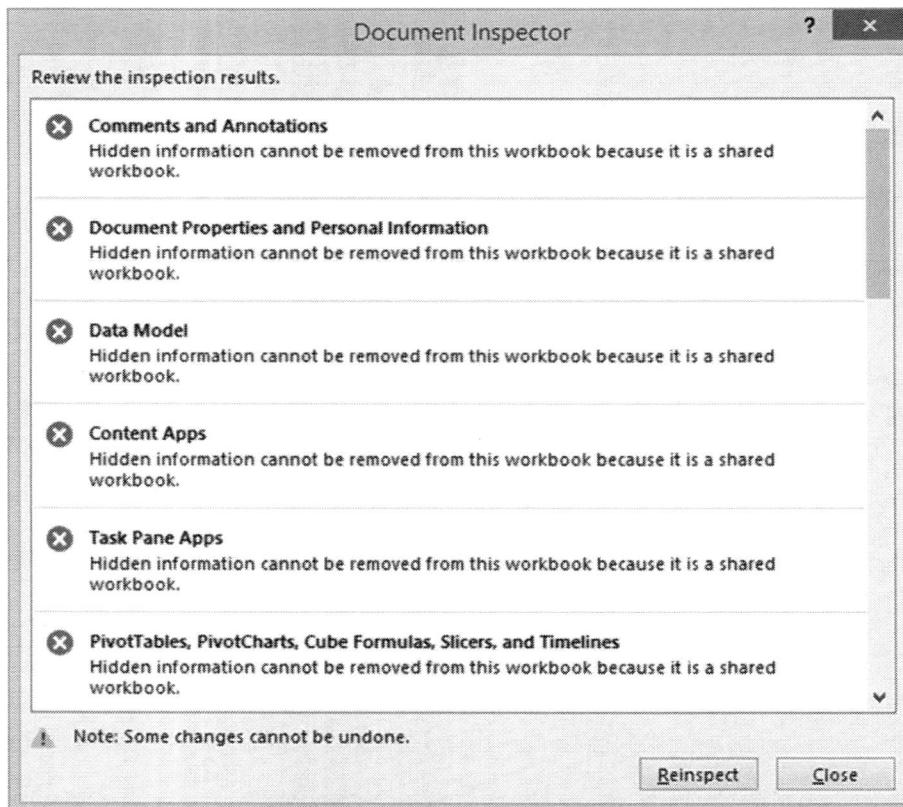

Document Inspector

Review the inspection results.

❌ **Comments and Annotations**
　　Hidden information cannot be removed from this workbook because it is a shared workbook.

❌ **Document Properties and Personal Information**
　　Hidden information cannot be removed from this workbook because it is a shared workbook.

❌ **Data Model**
　　Hidden information cannot be removed from this workbook because it is a shared workbook.

❌ **Content Apps**
　　Hidden information cannot be removed from this workbook because it is a shared workbook.

❌ **Task Pane Apps**
　　Hidden information cannot be removed from this workbook because it is a shared workbook.

❌ **PivotTables, PivotCharts, Cube Formulas, Slicers, and Timelines**
　　Hidden information cannot be removed from this workbook because it is a shared workbook.

⚠ Note: Some changes cannot be undone.

[Reinspect]　[Close]

Step 6: Select Remove All next to an item if you want to remove it. You can also return to the workbook and make the appropriate changes.

Step 7: Select Close when you have finished.

Sharing a Workbook

The following features are not supported in a shared workbook. You can put these items in (and save the workbook) BEFORE you share the workbook. You will not be able to make changes to those features after the workbook is shared.

You Can Not do this in a Shared Workbook	But You Can Do This
Create an Excel table	
Insert or delete blocks of cells	Insert entire rows and columns
Delete worksheets	
Merge cells or split merged cells	
Sort or filter by formatting	Sort or filter by number, text, or date, apply built-in filters, and filter by using the Search box
Add or change conditional formats	Use existing conditional formats as cell values change
Add or change data validation	Use data validation when you type new values
Create or change charts or PivotChart reports	View existing charts and reports
Insert or change pictures or other objects	View existing pictures and objects
Insert or change hyperlinks	Use existing hyperlinks
Use drawing tools	View existing drawings and graphics
Assign, change, or remove passwords	Use existing passwords

Protect or unprotect worksheets or the workbook	Use existing protection
Create, change, or view scenarios	
Use the Text to Columns command	
Group or outline data	Use existing outlines
Insert automatic subtotals	View existing subtotals
Create data tables	View existing data tables
Create or change PivotTable reports	View existing reports
Create or apply slicers	Existing slicers in a workbook are visible after the workbook is shared, but they cannot be changed for standalone slicers or be reapplied to PivotTable data or Cube functions. Any filtering that was applied for the slicer remains intact, whether the slicer is standalone or is used by PivotTable data or Cube functions in the shared workbook.
Create or modify spark lines	Existing spark lines in a workbook are displayed after the workbook is shared, and will change to reflect updated data. However, you cannot create new spark lines, change their data source, or modify their properties.
Write, record, change, view, or assign macros	Run existing macros that do not access unavailable features. You can also record shared workbook operations into a macro stored in another nonshared workbook.
Add or change Microsoft Excel 4 dialog sheets	

Change or delete array formulas	Excel will calculate existing array formulas correctly
Work with XML data	
Use a new data form to add new data.	Use a data form to find a record.

To add Sharing options to Ribbon

In earlier versions of Excel, there were several sharing options available on the review tab. In Microsoft

Excel 2016, these features are not turned on by default. To use the sharing option, you will need to add the features back onto the ribbon.

Step 1: Click the File ribbon and select Options.

Step 2: Click the Customize ribbon.

Step 3: Click the Review ribbon under the Customize the Ribbon menu on the right side of the page.

Step 4: Click New Tab button to add a menu to the Review ribbon.

Step 5: Click Rename to give the menu a contextual name. In the example, the menu is named "Share".

Step 6: Select All Commands from the list under Choose commands from drop-down list.

Step 7: Scroll down the list and select each of the following options and then click Add to add them to the menu on the ribbon.

- Share Workbook (Legacy)
- Track Changes (Legacy)
- Protect Sharing (Legacy)
- Compare and Merge Workbook (Legacy)

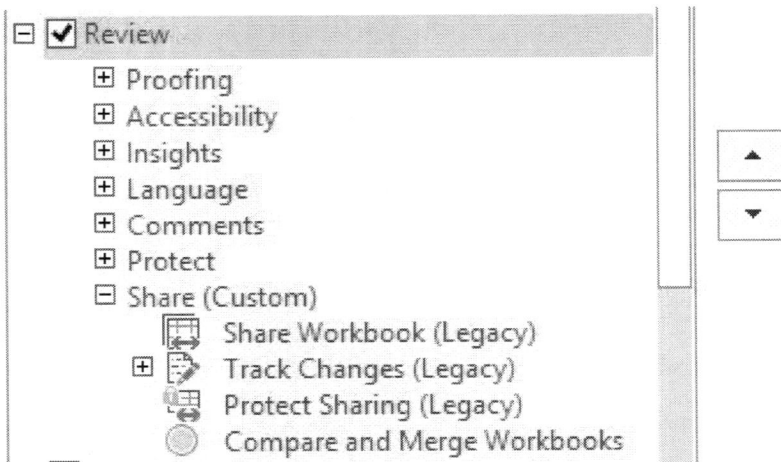

Step 8: Click Ok.

To share a workbook, use the following procedure

Step 1: Select the Review tab from the Ribbon.

Step 2: Select Share Workbook.

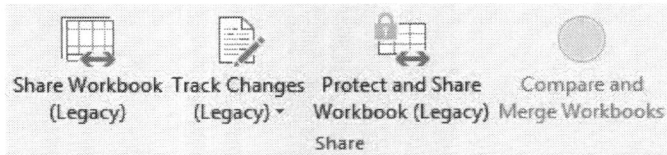

Step 3: In the Share Workbook dialog box, check the Allow changes by more than one user at the same time box.

Step 4: Select OK.

Step 5: In the Save As dialog box, enter a new File Name, if desired. Navigate to the location where you want to save the workbook and select Save.

Note that if there are links in the workbook, you may need to verify them.

Now you can email the people who will share the workbook a link to the file or indicate the location of the file.

Editing a Shared Workbook

To edit a shared workbook, use the following procedure.

Step 1: Open the shared workbook.

Step 2: Make any necessary changes.

Step 3: Save the workbook.

You can see who else has the workbook open on the Editing tab of the Share Workbook dialog box.

Step 1: Select the Review tab from the Ribbon.

Step 2: Select Share Workbook.

Select the Advanced tab of the Share Workbook dialog box to choose to get automatic updates of the other users' changes periodically, with or without saving.

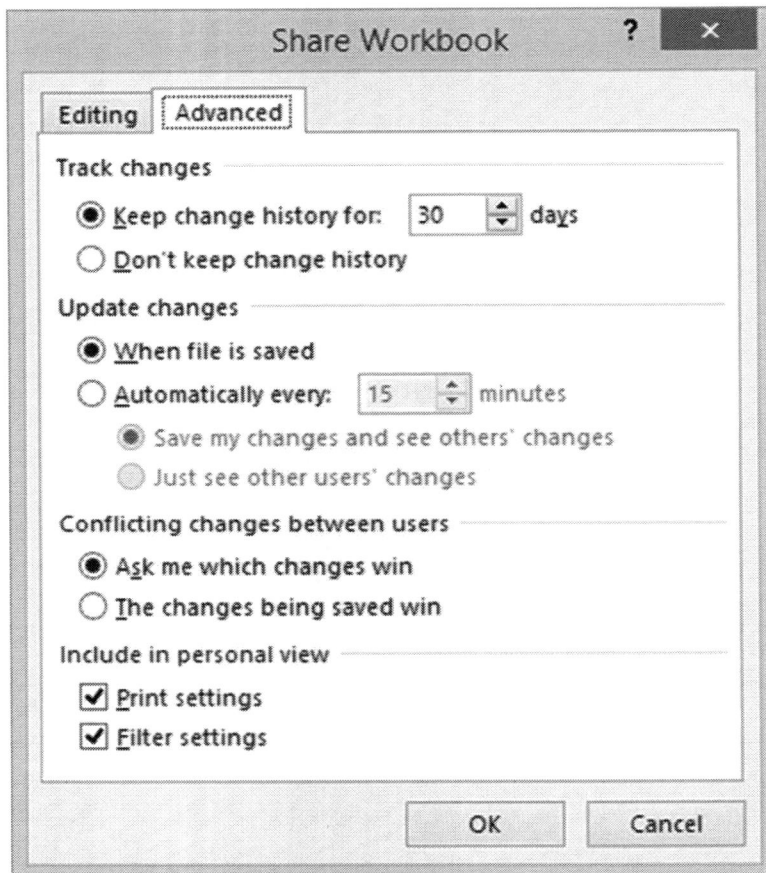

To turn on change tracking for a workbook, use the following procedure.

Step 1: Select the Review tab from the Ribbon.

Step 2: Select Share Workbook.

Step 3: Select the Advanced tab of the Share Workbook dialog box.

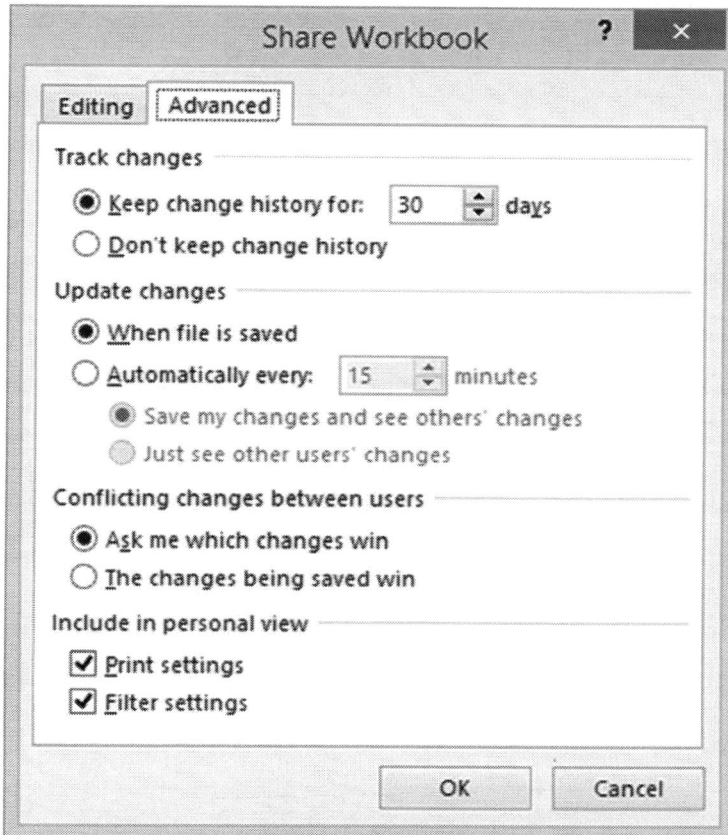

Step 4: Excel keeps the change history for a default of 30 days. You can change the number of days.

Step 5: Select OK.

To review the Highlight Changes dialog box, use the following procedure.

Step 1: Select the Review tab from the Ribbon.

Step 2: Select Track Changes. Select Highlight Changes.

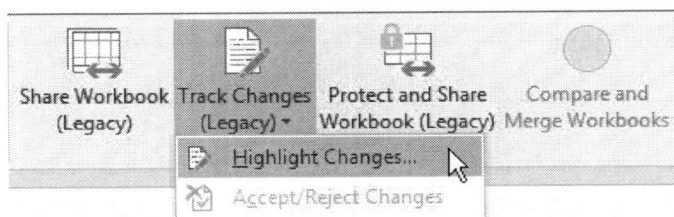

Step 3: In the Highlight Changes dialog box, you can select when to highlight changes, whose changes to highlight and where to highlight the changes. Select OK when you have finished making your selections.

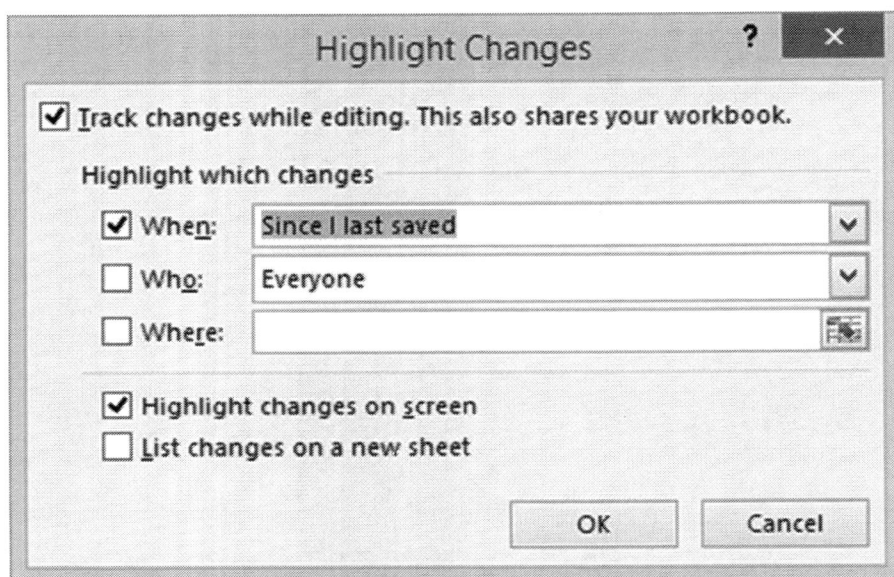

Merging Copies of a Shared Workbook

To compare and merge workbooks, use the following procedure.

Step 1: Open the copy of the shared workbook where you want to merge the changes.

Step 2: On the Review toolbar, click Compare and Merge Workbook (Legacy).

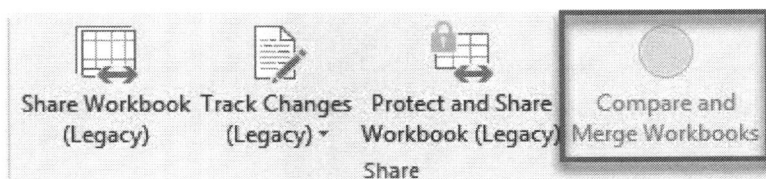

Step 3: Save the workbook if prompted.

Step 4: In the Select Files to Merge into Current Workbook dialog box, select a copy of the workbook that contains the changes that you want to merge. Hold down CTRL or SHIFT to select multiple copies. Select OK.

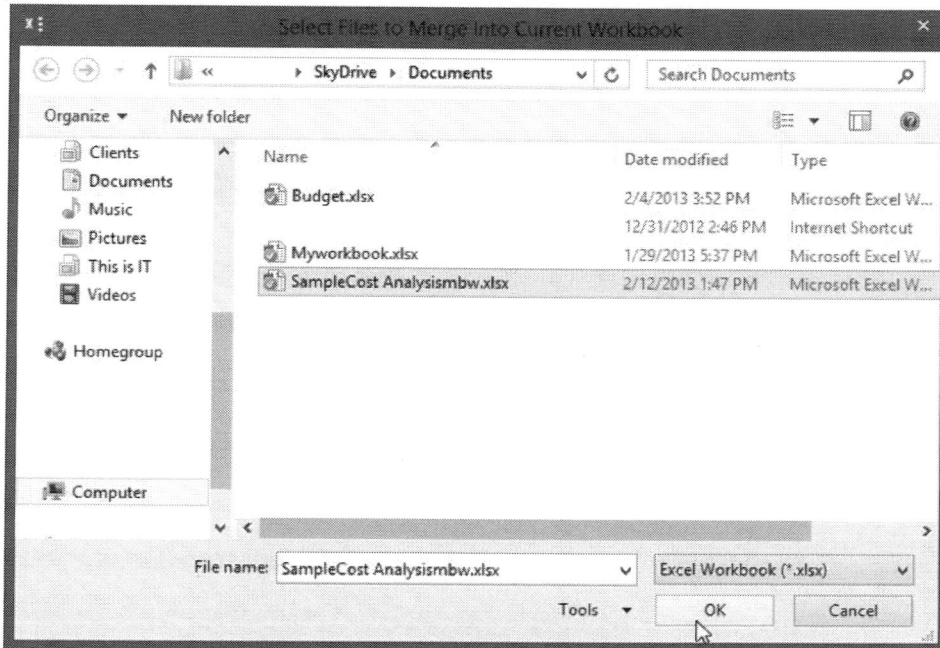

This chapter will help you with formulas and calculations. We will start with learning how to use the Watch Window, where you can monitor results of different areas of your workbook and even different workbooks related to the one you are changing. Then we will learn about Excel's methodology when calculating worksheets. You will learn how to set the calculation options for the current workbook and for all workbooks. This chapter also explains how to enable or disable automatic workbook calculations. Finally, we will look at the IFERROR function, which can help you evaluate formulas and display specific results if the formula contains an error.

Using the Watch Window

To use the Watch Window, use the following procedure.

Step 1: Select the Formulas tab from the Ribbon.

Step 2: Select Watch Window.

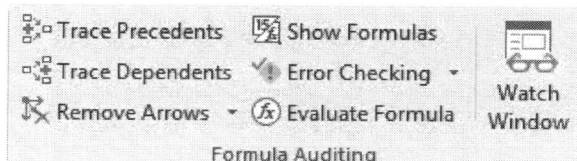

Step 3: In the Watch Window, select Add Watch.

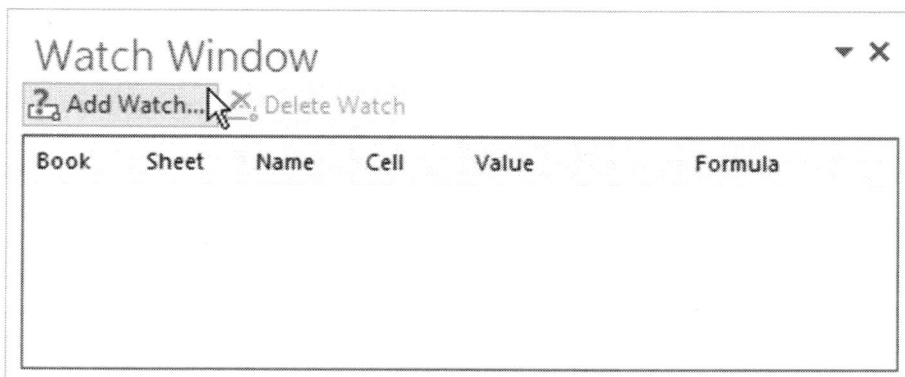

Step 4: In the Add Watch dialog box, indicate the cells that you want to watch. You can either type the cell references directly into the dialog box, or you can select the cells with your mouse.

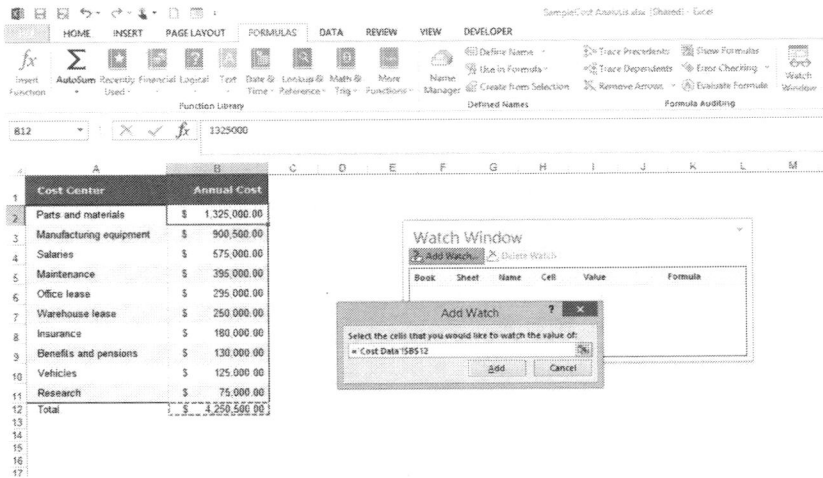

Note that you can select more than one contiguous cell. You can even select cells on another worksheet.

Step 5: Select Add.

Step 6: Notice that the Watch Window has started monitoring the selected cells. It indicates the workbook and worksheet where the cell(s) reside, the name of the cell (if you named it), the cell you are monitoring, the current value, and the formula you used to create that value.

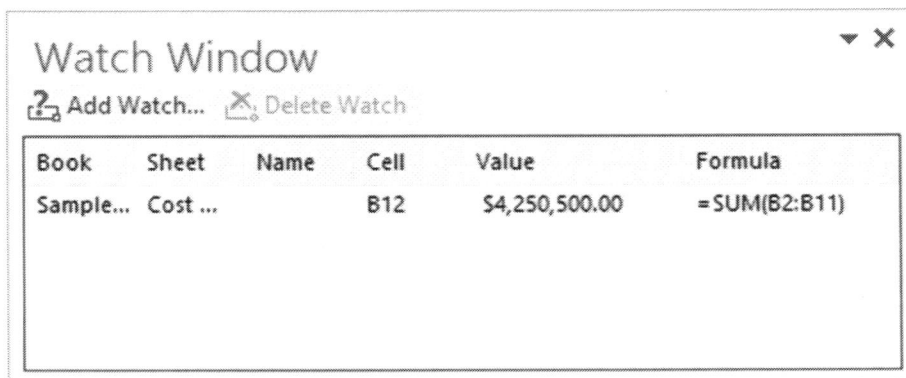

Book	Sheet	Name	Cell	Value	Formula
Sample...	Cost ...		B12	$4,250,500.00	=SUM(B2:B11)

Note that you can size the window by selected Size from the arrow in the top right corner. Then drag the corners to the new size.

You can also move the Watch Window to a more convenient location on your screen. Just drag the top border to the new location. In this example, it is docked to the top of the screen, just below the Ribbon. You can also dock it to the left or the right side of the screen.

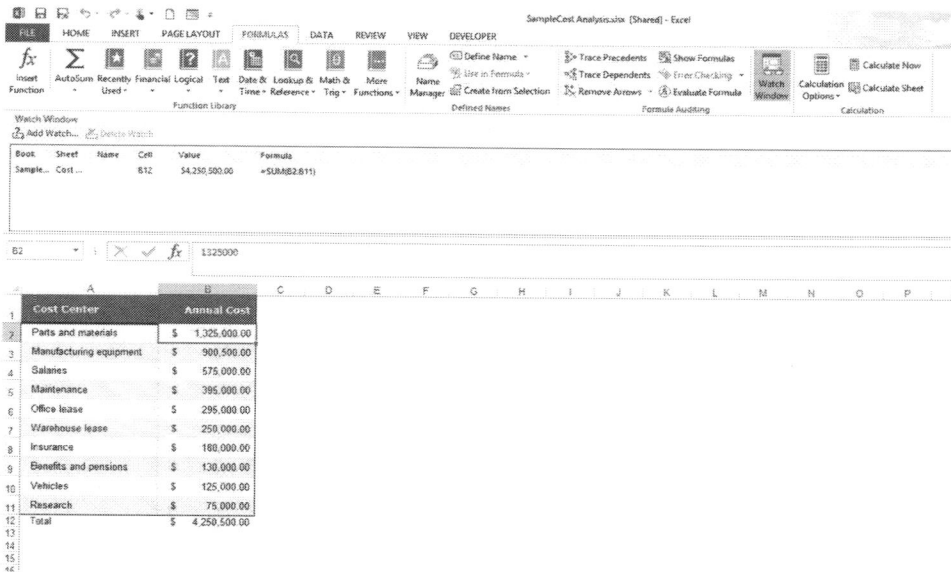

Add more areas to watch and make changes to the watched cells to see the results.

To delete a watch, highlight the item in the Watch Window you no longer need and select Delete Watch.

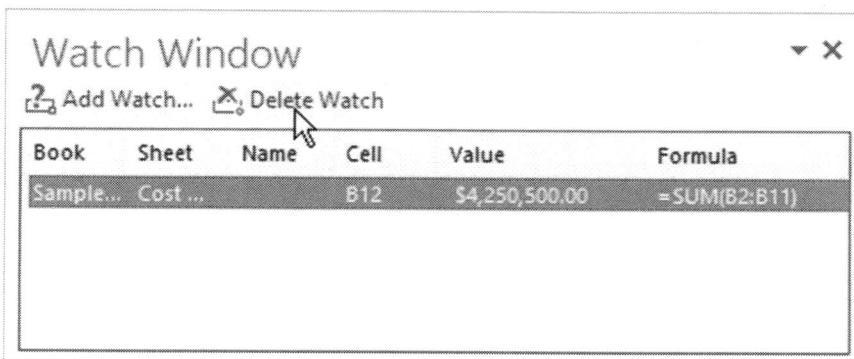

About the Excel Calculation Process

The three stages in the process of calculating data in Excel are:

- Constructing a dependency tree
- Constructing a calculation chain

- Recalculating cells

Dependency Tree and Calculation Chain

Excel uses the dependency tree to determine precedents. Excel uses this dependency tree to construct a calculation chain, which lists all cells that contain formulas in the order in which they should be calculated.

Each time the workbook is recalculated, Excel revises the Calculation chain if it comes across a formula that depends on a cell that has not been calculated. This means that calculation times improve from the first time the worksheet has been opened to the completion of the first few calculation cycles.

Dirty Cells

Anytime you make a structural change to a workbook (including entering a new formula), Excel must reconstruct the dependency tree and calculation train. Excel marks all cells that depend on a new formula or cell of data that you enter as dirty. Direct and indirect dependents are marked as dirty. At this time, Excel also detects any circular references and displays a warning message.

Excel re-evaluates the contents of each dirty cell in the order dictated by the calculation chain. With automatic calculations, the recalculation for these cells happens immediately after marking the cells as dirty.

Setting Calculation Options

Use the following procedure to set calculation options at the workbook level.

Step 1: Select the Formulas tab from the Ribbon.

Step 2: Select Calculation Options.

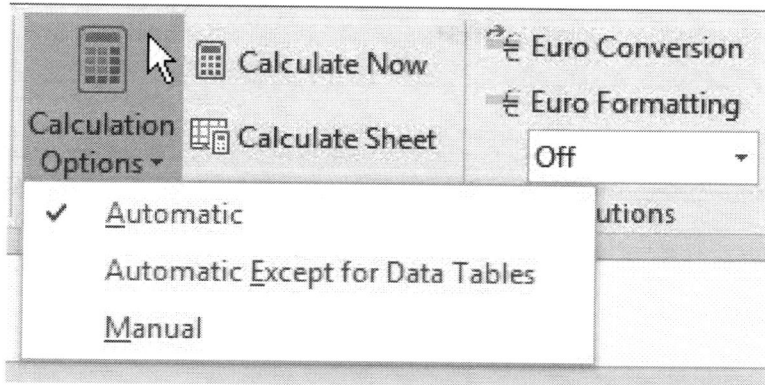

Step 3: Select one of the calculation options.

Step 4: If you do not use Automatic calculation, you can use the Calculate Now and Calculate Sheet commands in the same group on the Formulas tab, using the Customize the Ribbon tab on the Excel Options dialog box.

Enabling or Disabling Automatic Workbook Calculations

To use the Options dialog box to set the Calculation Options, use the following procedure.

Note that setting Calculation options in this way applies to the application level rather than the workbook level. That means it applies for all open workbooks.

Step 1: Select the File tab from the Ribbon.

Step 2: Select Options.

Step 3: Select Formulas.

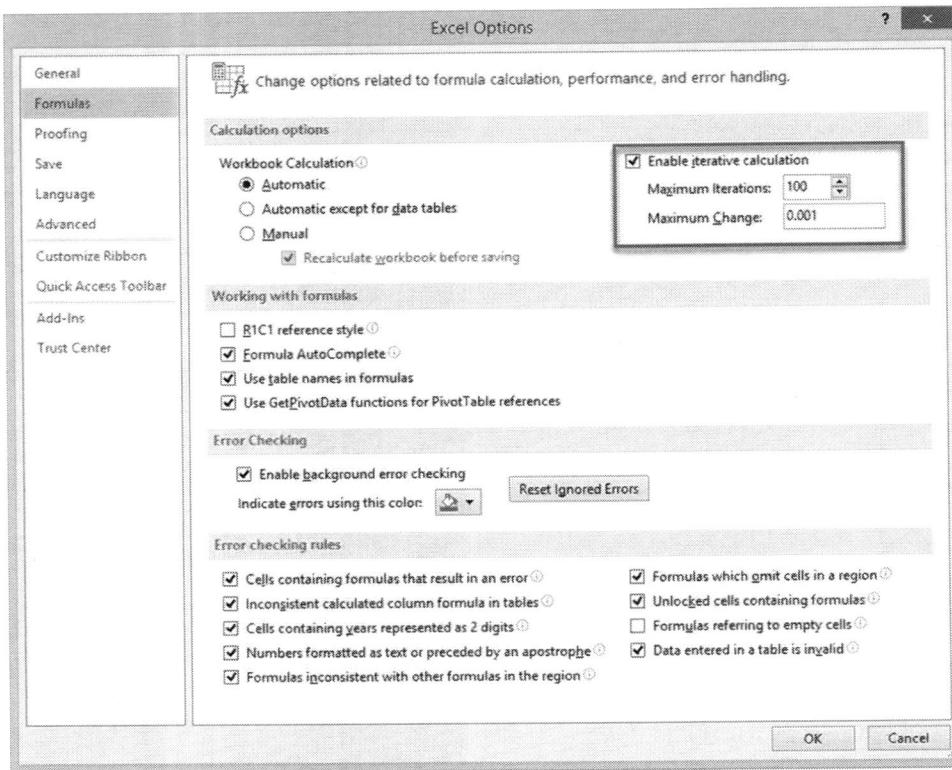

You can also set Iterative Calculation in the Options dialog box. Check the box and select several Maximum Iterations. Also indicate the Maximum change amount.

Note that the Iterative calculation options are a session-based option. That means that next time you open Excel, you will need to reset the Iterative calculation options.

Using the IFERROR Function to Evaluate Formulas

To use the IFERROR function, use the following procedure.

Step 1: In the cell that includes the formula you want to check, enter =IFERROR. When you start typing, the IFERROR function will display in a drop-down menu. You can double-click to select it.

Step 2: Enter an open parenthesis, then the formula that you want to use.

Step 3: Enter a comma after the formula, then the value to display if the formula returns an error. Remember that if you want to use text, it must be in quotation marks.

Step 4: Now enter a close parenthesis and press Enter.

Step 5: Notice that when you fill the new formula into all the cells in Column K, the division by zero errors has been replaced with the words "No Contract."

Chapter 24 – Working with Array Formulas

This chapter introduces using array formulas. We will first look at what array formulas are and some advantages, disadvantages, and rules when using them, as well as an introduction to array constants. Then you will practice creating simple and more advanced arrays.

About Array Formulas

Array formulas allow you to do complex tasks in Excel. For example, you may want to count the number of characters in a range of cells or sum only numbers that meet certain conditions. Array formulas are also known as CSE formulas because of the keystrokes you use to enter the formula (CTRL + SHIFT + ENTER).

An array is a collection of items that, in Excel, can reside in a single row, a single column or multiple rows and columns. A single-row or column array is known as a one-dimensional array (either one-dimensional horizontal or one-dimensional vertical). Multiple row and column arrays are known as two-dimensional arrays. Excel does not allow you to create three-dimensional array formulas.

An array formula can perform multiple calculations on one or more items in the array. You can return either multiple results or a single result.

Array formulas offer several advantages. They provided consistency in your workbook, which helps ensure greater accuracy. They also provide safety from accidentally overwriting a component of the array. Changes to array formulas must be confirmed by pressing CTRL + SHIFT + ENTER. Another advantage is that using array formulas result in smaller file sizes than using several intermediate formulas.

Array formulas also have some disadvantages. Array formulas are typically undocumented in a worksheet, which may cause problems if other users do not understand how to use array formulas. Also, large array formulas can slow down calculations.

Rules for entering and changing multi-cell array formulas

In addition to pressing CTRL+SHIFT+ENTER whenever you need to enter or edit an array formula, multi-cell formulas require some additional rules.

- Select the range of cells to hold your results before you enter the formula.
- You cannot change the contents of an individual cell in an array formula.
- You can move or delete an entire array formula, but you cannot move or delete part of it. In other words, to shrink an array formula, you first delete the existing formula and then start over.
- You cannot insert blank cells into or delete cells from a multi-cell array formula.

Array Constants

You use array constants as a component of array formulas. To create an array constant, you enter a list of items and surround the list with braces. For example, here is a simple array constant: ={1,2,3,4,5}

A horizontal array is created by separating the constant list items with a comma. You can create a vertical array by using semicolons instead. A two-dimensional array includes definitions for both rows and columns, so the list will use both commas and semicolons.

Constants can include numbers, text, logical values and error values. Numbers can appear in integer, decimal, and scientific formats. Remember to include text in quotation marks (""). Array constants cannot include additional arrays, formulas, or functions.

Creating One-Dimensional and Two-Dimensional Constants
To create a horizontal constant, use the following procedure.

Step 1: In the first row of the blank workbook, enter the following information:

▲	A	B	C	D	E	F
1	1	2	3	4	5	
2						

Step 2: Select cells A1 through E1.

220

Step 3: In the formula bar, enter the following formula

={1,2,3,4,5}

Step 4: Press CTRL + SHIFT + ENTER.

Excel enters the constant into each cell that you selected.

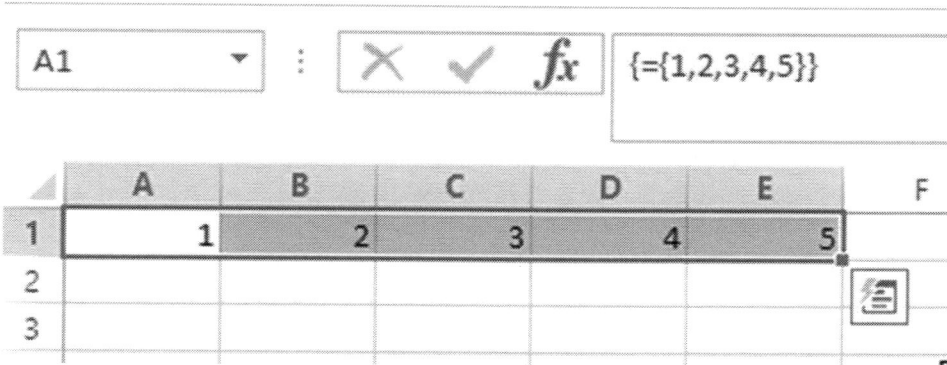

A1				{={1,2,3,4,5}}
	A	B	C	D	E	F
1	1	2	3	4	5	
2						
3						

To create a vertical constant, use the following procedure.

Step 1: In the first column of a blank worksheet, enter the following into cells A1 through A5:

	A
1	1
2	2
3	3
4	4
5	5
6	

Step 2: Select cells A1 through A5.

Step 3: In the formula bar, enter the following formula

={1;2;3;4;5}

Step 4: Press CTRL + SHIFT + ENTER.

Again, Excel enters the constant into each selected cell.

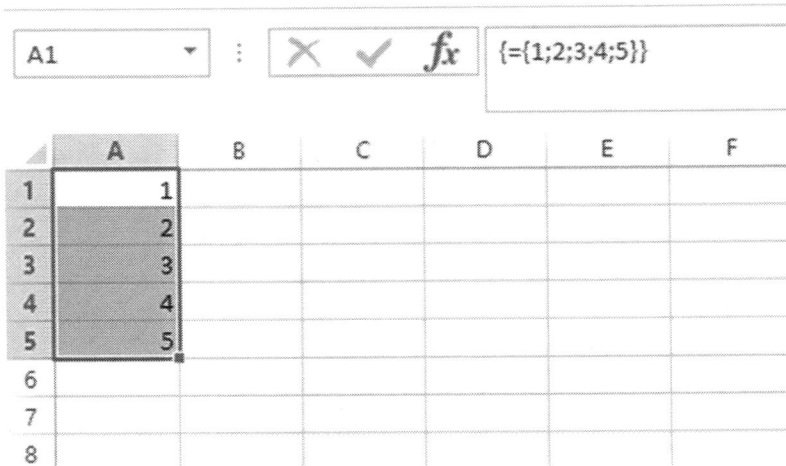

| A1 | | : | X | ✓ | *fx* | {={1;2;3;4;5}} |

	A	B	C	D	E	F
1	1					
2	2					
3	3					
4	4					
5	5					
6						
7						
8						

To create a two-dimensional constant, use the following procedure.

Step 1: In a blank worksheet, enter a number from 1 to 12 in each of the first four rows and first four columns, as below:

	A	B	C	D
1	1	2	3	4
2	5	6	7	8
3	9	10	11	12

Step 2: Select cells A1 through D3.

Step 3: In the formula bar, enter the following formula

={1,2,3,4;5,6,7,8;9,10,11,12}

Step 4: Press CTRL + SHIFT + ENTER.

Notice the results.

| A1 | ▾ | : | ✕ ✓ *fx* | {={1,2,3,4;5,6,7,8;9,10,11,12}} |

	A	B	C	D	E	F	G
1	1	2	3	4			
2	5	6	7	8			
3	9	10	11	12			
4							
5							
6							
7							

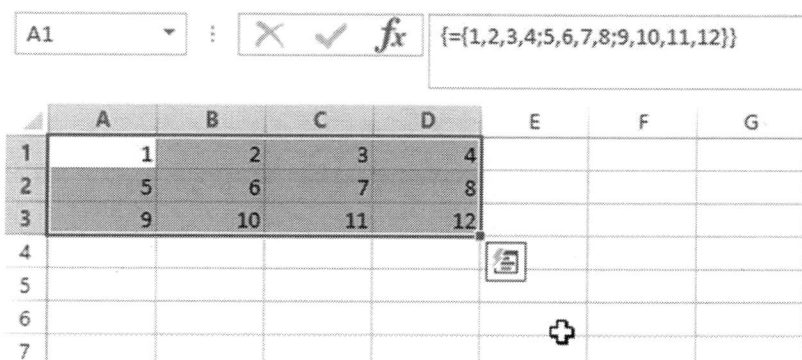

To use an array constant in a formula, use the following procedure.

Step 1: On the sheet from the previous procedures that includes the horizontal constant, copy the following formula in cell A3:

=SUM(A1:E1*{1,2,3,4,5})

Step 2: Press CTRL + SHIFT + ENTER.

Notice that Excel adds brackets around the entire formula. This formula multiplies the values stored in the stored array by the corresponding values in the constant. It is the equivalent of =SUM(A1*1,B1*2,C1*3,D1*4,E1*5).

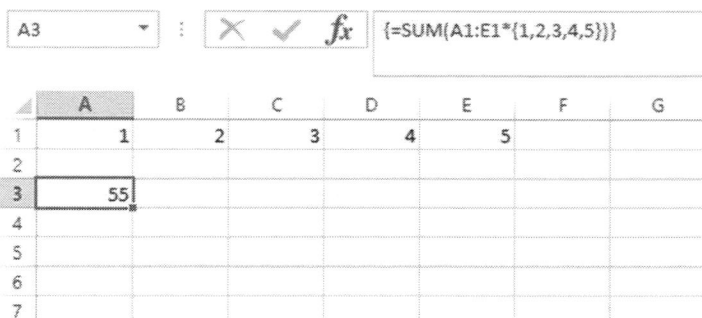

| A3 | ▾ | : | ✕ ✓ *fx* | {=SUM(A1:E1*{1,2,3,4,5})} |

	A	B	C	D	E	F	G
1	1	2	3	4	5		
2							
3	55						
4							
5							
6							
7							

To name an array constant, use the following procedure.

Step 1: Select the Formulas tab.

Step 2: Select Define Name.

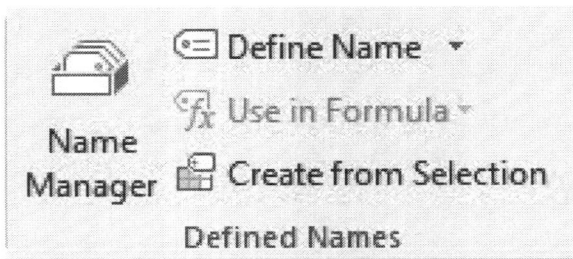

Step 3: In the New Name dialog box, enter the name. In this example, we will call it Quarter1.

Step 4: In the Refers to box, enter the following constant. Remember to type the braces manually.

={"January","February","March"}

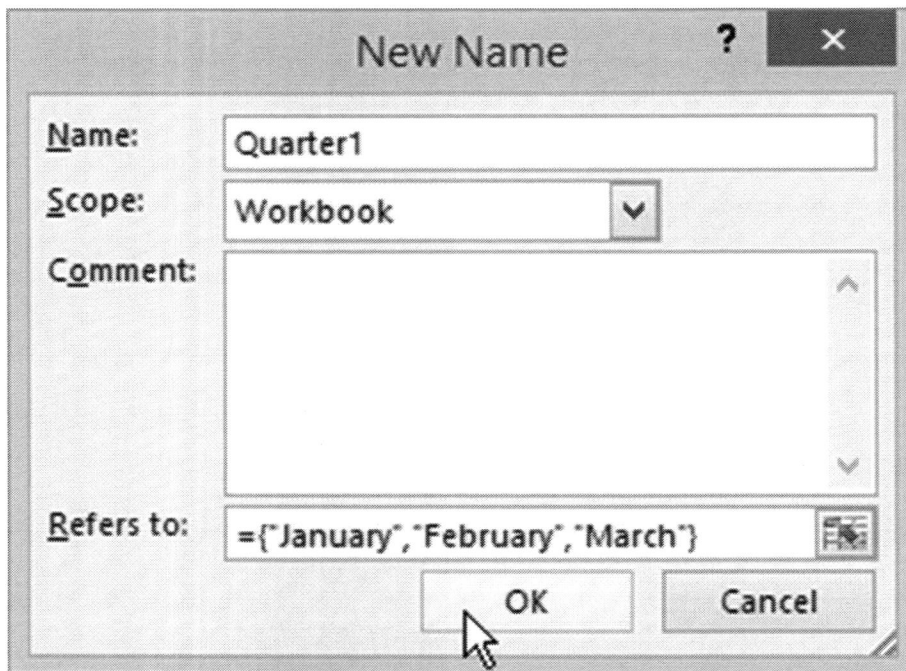

Step 5: Select OK.

Now let us try using the named array constant.

Step 1: Select a row of three blank cells in the worksheet.

Step 2: Enter the following formula:

=Quarter1

Step 3: Press CTRL + SHIFT + ENTER.

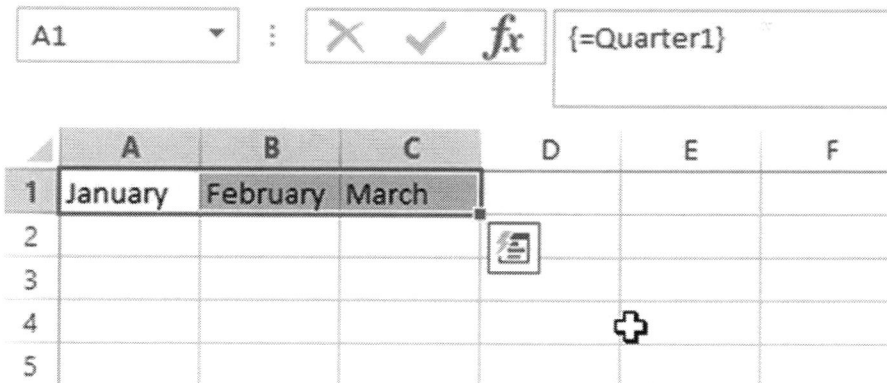

| A1 | ▾ | : | ✕ | ✓ | fx | {=Quarter1} |

◢	A	B	C	D	E	F
1	January	February	March			
2						
3						
4						
5						

Creating a Simple Array

To create an array from existing values.

Step 1: In the sample workbook, on the Arrays worksheet, select cells C1 through E3.

Step 2: Enter the following formula in the formula bar:

=Data!E1:G3

Step 3: Press CTRL + SHIFT + ENTER.

The following result is displayed.

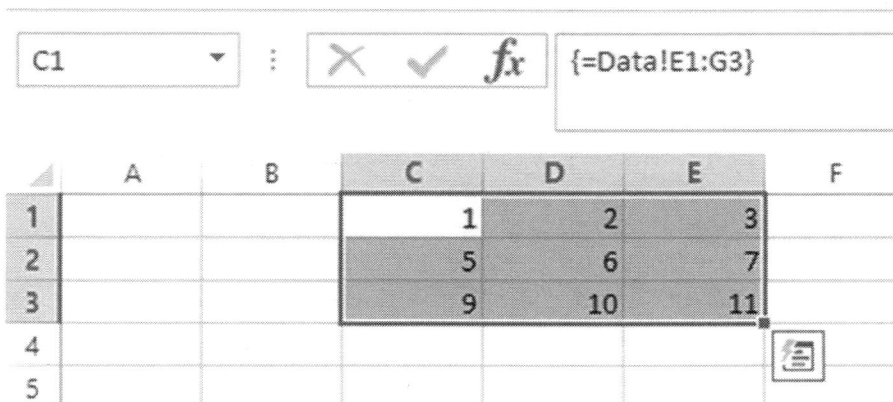

| C1 | ▾ | : | ✕ | ✓ | fx | {=Data!E1:G3} |

◢	A	B	C	D	E	F
1			1	2	3	
2			5	6	7	
3			9	10	11	
4						
5						

The formula links to the values stored in cells E1 through G3 on the Data worksheet.

(You could also accomplish this by putting a separate, unique formula in each cell of the Arrays worksheet).

If you change some of the values on the Data worksheet, those changes appear on the Arrays worksheet. Remember that you will have to follow the Array rules if you need to change any of the data.

To create an array constant from existing values, use the following procedure.

Step 1: In the sample workbook, on the Arrays worksheet, select cells C1 through E3.

Step 2: Press F2 to switch to edit mode.

Step 3: Press F9 to convert the cell references to values.

Excel converts the values into an array constant.

Step 4: Press CTRL+SHIFT+ENTER to enter the array constant as an array formula.

Step 5: Excel replaces the =Data!E1:G3 array formula with the following array constant:

={1,2,3;5,6,7;9,10,11}

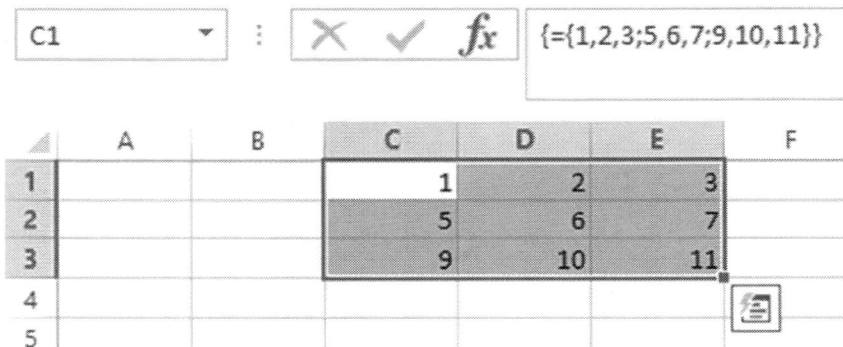

This means that the link between the Data and Arrays worksheets is no longer there. The Array constant replaces the array formula.

To create an array that counts the characters in a range of cells, use the following procedure.

Step 1: On the Data worksheet, enter the following formula in cell C7:

=SUM(LEN(C1:C5))

Step 2: Press CTRL + SHIFT + ENTER.

The LEN function returns the length of each text string in the range. The SUM function then adds those values together and displays the result in the cell with the formula.

C7		:	X ✓ *fx*	{=SUM(LEN(C1:C5))}						

	A	B	C	D	E	F	G	H	I	J	K
1	400		the quick		1	2	3	4			
2	1200		brown fox		5	6	7	8			
3	3200		jumped over		9	10	11	12			
4	475		the lazy		13	14	15	16			
5	500		power user								
6	2000										
7	600		47								
8	1700										
9	800										
10	2700										
11											

To create an array that finds the N smallest values in a range, use the following procedure.

Step 1: On the Data worksheet, select cells A12 through A14 to hold the results returned by the array formula.

Step 2: Enter the following formula:

=SMALL(A1:A10,{1;2;3})
Step 3: Press CTRL + SHIFT + ENTER.

The SMALL function is evaluated three times in this formula. The result is the first, second, and third smallest members of the array contained in cells A1:A10. You could add more arguments to the constant to find more values. You could replace the SMALL function with the LARGE function to find the largest values in a range.

	A	B	C	D	E	F	G	H	I
A12					fx	{=SMALL(A1:A10,{1;2;3})}			

	A	B	C	D	E	F	G	H	I
1	400		the quick		1	2	3	4	
2	1200		brown fox		5	6	7	8	
3	3200		jumped over		9	10	11	12	
4	475		the lazy		13	14	15	16	
5	500		power user						
6	2000								
7	600		47						
8	1700								
9	800								
10	2700								
11									
12	400								
13	475								
14	500								
15									
16									
17									

You could also use the SUM or AVERAGE functions with this formula, such as

=SUM(SMALL(A1:A10,{1;2;3}))

=AVERAGE(SMALL(A1:A10,{1;2;3;})

To create an array to find the longest text string in a range of cells, use the following procedure.

Step 1: On the Data worksheet, delete the contents of C7.

Step 2: Enter the following formula:

=INDEX(C1:C5,MATCH(MAX(LEN(C1:C5)),LEN(C1:C5),0),1)
Step 3: Press CTRL + SHIFT + ENTER.

Let us examine this formula, starting with the inner elements. The LEN function returns the length of each item, as shown before. The MAX function calculates the largest value among those items. Then the MATCH function calculates the offset, or relative position, of the cell that contains the longest text string. It requires three arguments, which are a lookup value, a lookup array, and a match type. The MATCH type argument is 0 in this case. Finally, the INDEX function has an array and a row and column number within that array as its arguments. The MATCH function provides the cell address.

228

C7 | {=INDEX(C1:C5,MATCH(MAX(LEN(C1:C5)),LEN(C1:C5),0),1)}

	A	B	C	D	E	F	G	H	I	J
1	400		the quick		1	2	3	4		
2	1200		brown fox		5	6	7	8		
3	3200		jumped over		9	10	11	12		
4	475		the lazy		13	14	15	16		
5	500		power user							
6	2000									
7	600		jumped over							
8	1700									
9	800									
10	2700									
11										
12	400									
13	475									
14	500									
15										

Creating an Advanced Array

The SUM function in Excel does not work when you try to sum a range that contains an error value, such as #N/A.

To create an array that sums a range that contains error values, use the following procedure.

Step 1: In Cell K35, enter the following formula:

=SUM(IF(ISERROR(K3:K33),"",(K3:K33)))

Step 2: Press CTRL + SHIFT + ENTER.

This formula creates a new array that includes the original values, but does not include the error values.

K35 | {=SUM(IF(ISERROR(K3:K33),"",(K3:K33)))}

	C	D	E	F	G	H	I	J	K	L	M
22	2	2	2	5	2	3	3	3	$7.70		
23	3	4	3	2	2	2	2	4	$16.40		
24	1	2	0	1	1	2	1	1	#DIV/0!		
25	3	2	4	5	4	3	2	2	$13.14		
26	4	1	4	3	5	2	5	5	$11.00		
27	3	4	5	3	2	0	5	4	$10.67		
28	3	5	3	4	5	2	1	4	$11.22		
29	2	5	4	5	5	3	5	4	$15.38		
30	3	0	5	5	1	4	2	3	$40.00		
31	4	3	5	1	1	1	0	5	$18.50		
32	2	4	2	5	5	2	4	1	$94.00		
33	1	1	4	3	3	5	4	4	$18.60		
34											
35										$672.04	
36											

To create an array that counts the number of error values in a range, use the following procedure.

Step 1: In Cell K37, enter the following formula:

=SUM(IF(ISERROR(K3:K33),1,0))

Step 2: Press CTRL + SHIFT + ENTER.

This formula is similar to the previous one, but instead of omitting the error cells from the sum, it returns the number of error values in the range.

	C	D	E	F	G	H	I	J	K	L
							fx		{=SUM(IF(ISERROR(K3:K33),1,0))}	
22	2	2	2	5	2	3	3	3	$7.70	
23	3	4	3	2	2	2	2	4	$16.40	
24	1	2	0	1	1	2	1	1	#DIV/0!	
25	3	2	4	5	4	3	2	2	$13.14	
26	4	1	4	3	5	2	5	5	$11.00	
27	3	4	5	3	2	0	5	4	$10.67	
28	3	5	3	4	5	2	1	4	$11.22	
29	2	5	4	5	5	3	5	4	$15.38	
30	3	0	5	5	1	4	2	3	$40.00	
31	4	3	5	1	1	1	0	5	$18.50	
32	2	4	2	5	5	2	4	1	$94.00	
33	1	1	4	3	3	5	4	4	$18.60	
34										
35									$672.04	
36										
37									2	
38										

In the next example, let us imagine that you need to sum just the positive integers in a range named Sales, which is already defined on Sheet 2.

To create an array that sums values based on conditions, use the following procedure.

Step 1: In cell K 39, enter the following formula:

=SUM(IF(Sales>0,Sales))

Step 2: Press CTRL + SHIFT + ENTER.

The IF function creates an array of positive values and false values. The SUM function ignores the false values.

230

K39 | fx {=SUM(IF(Sales>0,Sales))}

	C	D	E	F	G	H	I	J	K	L
22	2	2	2	5	2	3	3	3	$7.70	
23	3	4	3	2	2	2	2	4	$16.40	
24	1	2	0	1	1	2	1	1	#DIV/0!	
25	3	2	4	5	4	3	2	2	$13.14	
26	4	1	4	3	5	2	5	5	$11.00	
27	3	4	5	3	2	0	5	4	$10.67	
28	3	5	3	4	5	2	1	4	$11.22	
29	2	5	4	5	5	3	5	4	$15.38	
30	3	0	5	5	1	4	2	3	$40.00	
31	4	3	5	1	1	1	0	5	$18.50	
32	2	4	2	5	5	2	4	1	$94.00	
33	1	1	4	3	3	5	4	4	$18.60	
34										
35									$672.04	
36										
37									2	
38										
39									603.6595	
40										
41										

To create an array that computes an average that excludes zeros, use the following procedure.

Step 1: In cell D35, enter the following formula:

=AVERAGE(IF((D3:D33)<>0,(D3:D33)))

Step 2: Press CTRL + SHIFT + ENTER.

K39 | fx {=SUM(IF(Sales>0,Sales))}

	C	D	E	F	G	H	I	J	K	L
22	2	2	2	5	2	3	3	3	$7.70	
23	3	4	3	2	2	2	2	4	$16.40	
24	1	2	0	1	1	2	1	1	#DIV/0!	
25	3	2	4	5	4	3	2	2	$13.14	
26	4	1	4	3	5	2	5	5	$11.00	
27	3	4	5	3	2	0	5	4	$10.67	
28	3	5	3	4	5	2	1	4	$11.22	
29	2	5	4	5	5	3	5	4	$15.38	
30	3	0	5	5	1	4	2	3	$40.00	
31	4	3	5	1	1	1	0	5	$18.50	
32	2	4	2	5	5	2	4	1	$94.00	
33	1	1	4	3	3	5	4	4	$18.60	
34										
35									$672.04	
36										
37									2	
38										
39									603.6595	
40										
41										

Chapter 25 – Working with Macros

In this chapter, you will learn how to assign a macro you have created to a command button, which you can easily access each time you want to run the macro. You will also learn how to set up a graphical area that causes a macro to run when it is clicked. Similarly, this chapter explains how to run a macro automatically when a spreadsheet is opened. Finally, you will learn how to change a macro.

Assigning a Macro to a Command Button

To assign a new macro to a command key, use the following procedure.

Step 1: Select the File tab from the Ribbon to open the Backstage view.

Step 2: Select the Options tab on the left.

Step 3: Select Customize Ribbon.

Step 4: In the Choose commands from drop down list, select Macros.

You will need to create a Custom Group on the Ribbon before you can assign a macro to a Ribbon tab.

Step 1: Select New Group.

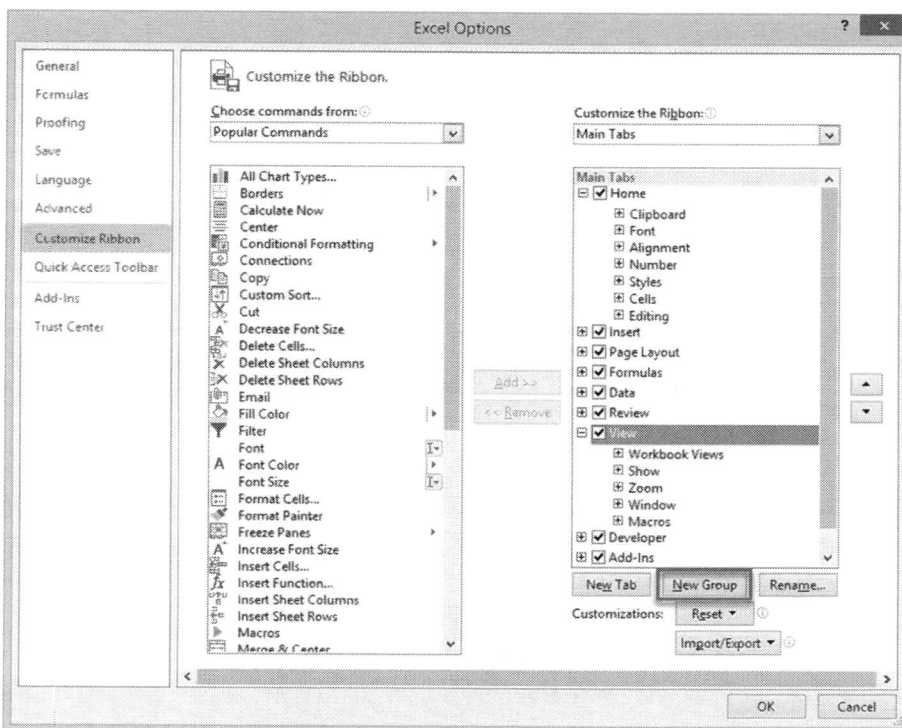

Step 2: Select Rename.

Step 3: Enter a new Display name.

Step 4: Select OK.

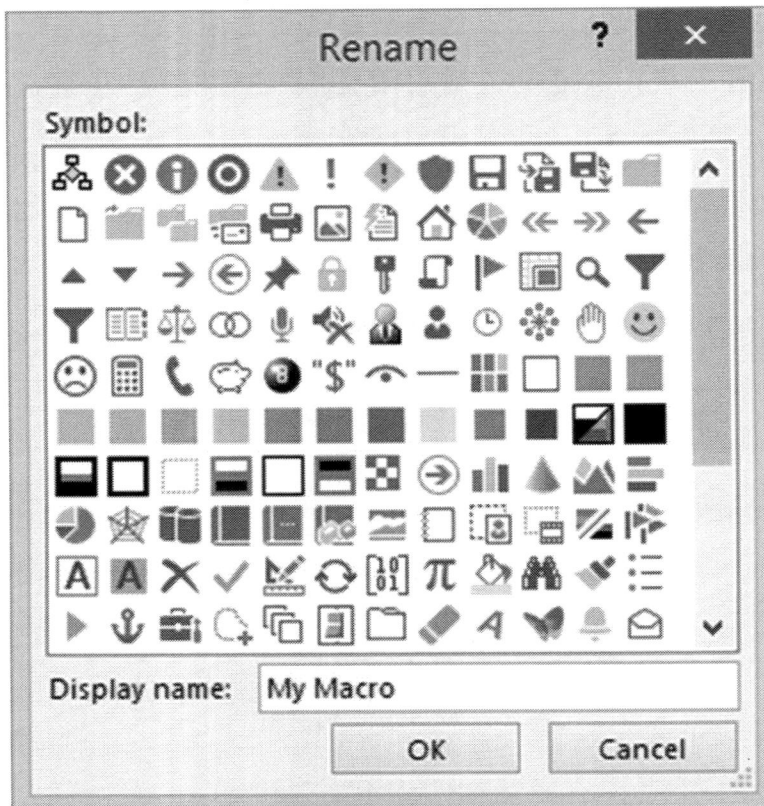

Now add the macro to the group.

Step 1: In the Customize the Ribbon list, select the ribbon where you would like to display the Macro command button.

Step 2: In the Choose Command from list, select your macro.

Step 3: Select Add.

Step 4: Select Rename.

Step 5: Select an icon for the macro from the list of Symbols.

Step 6: Select OK.

Now look at the selected Ribbon and see the macro command that you added.

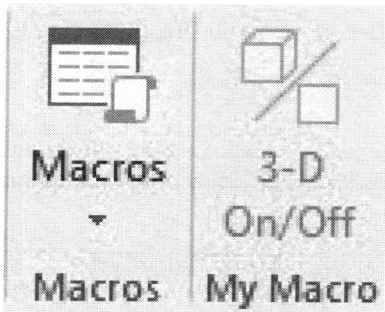

Running a Macro by Clicking an Area of a Graphic Object

To assign a macro to a graphical object, use the following procedure.

Step 1: Insert a shape. To do this, select the Insert tab from the Ribbon. Select Shapes. Select the shape you want to insert.

Step 2: Right-click on the shape.

Step 3: Select Assign Macro from the context menu.

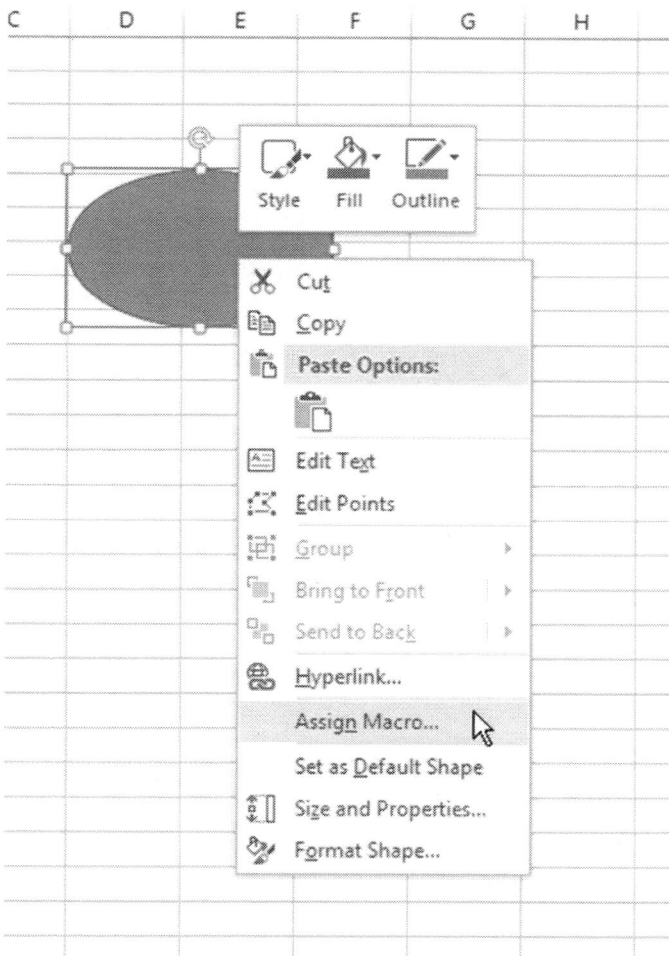

Step 4: Select the Macro name from the list. You can choose the location where the Macro is stored by selecting a new option from the Macros in drop down list.

Step 5: Select OK.

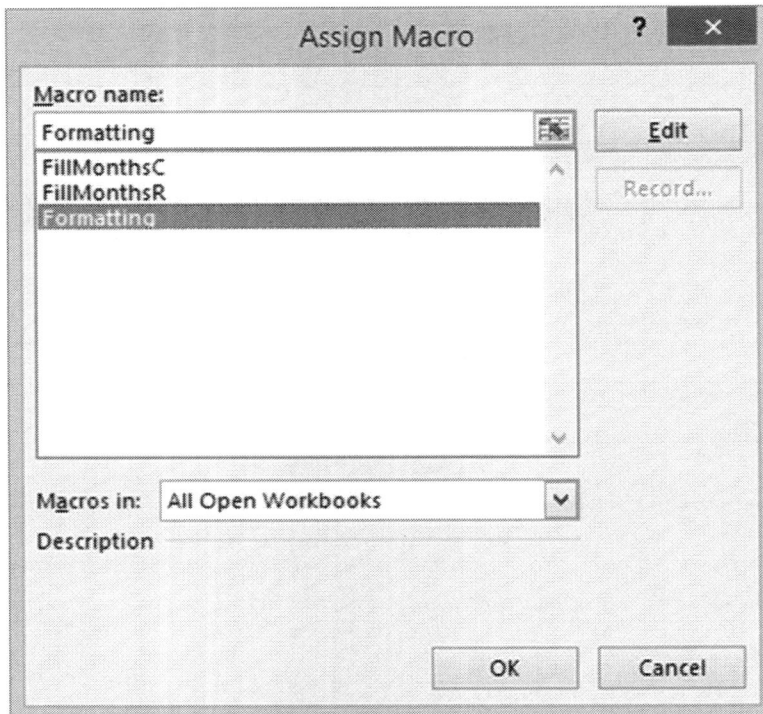

Now try running the macro by clicking on the object.

You can format the shape in any way desired.

Configuring a Macro to Run Automatically Upon Opening the Workbook

To create an Auto_Open macro, use the following procedure.

Step 1: Select the View tab from the Ribbon.

Step 2: Select Macros.

Step 3: Select Record Macro.

Step 4: In the Macro name box, enter Auto_Open as the name.

Step 5: In the Store macro in list, select the workbook where you want to store the macro from the drop-down list.

Step 6: Select OK.

Step 7: Perform the actions that you want the macro to perform. For a simple example, simply select Zoom to Selection from the View tab.

Step 8: Select Macros from the View Tab. Select Stop Recording.

Step 9: Save the workbook. You will need to select Excel macro-Enabled Workbook (*.xlsm) from the Save as type drop down list.

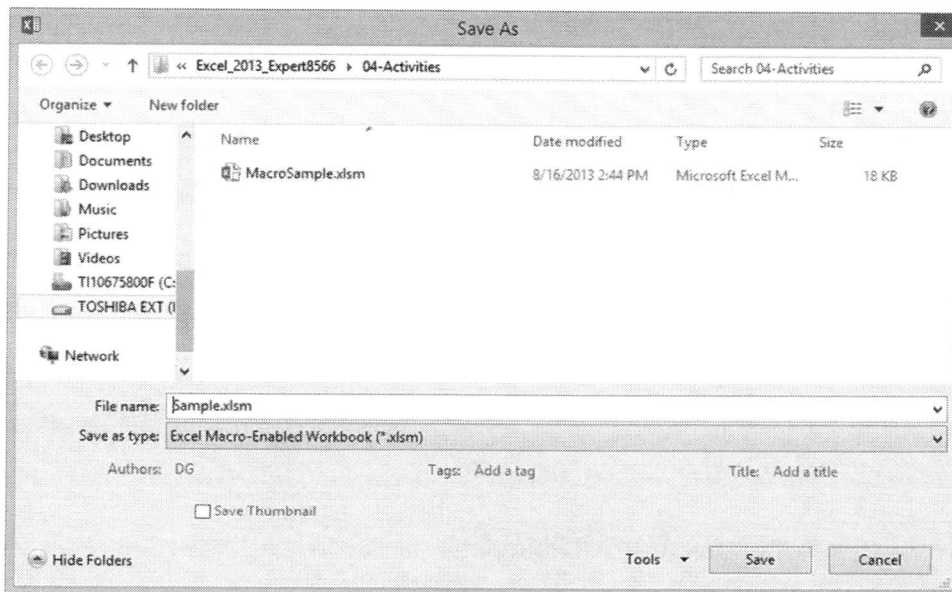

To test out your auto macro, close the workbook and reopen it. The macro is performed as soon as you open the workbook.

Recording an Auto_Open macro has the following limitations:

- If the workbook where you save the Auto_Open macro already contains a VBA procedure in its Open event, the VBA procedure for the Open event will override all actions in the Auto_Open macro.

- An Auto_Open macro is ignored when a workbook is opened programmatically by using the Open method.

- An Auto_Open macro runs before any other workbooks open. Therefore, if you record actions that you want Excel to perform on the default Book1 workbook or on a workbook that is loaded from the XLStart folder, the Auto_Open macro will fail when you restart Excel, because the macro runs before the default and startup workbooks open.

To start a workbook without running an Auto_Open macro, hold down the SHIFT key when you start Excel.

Changing a Macro

To change the name of a macro, use the following procedure.

Step 1: Select the View tab from the Ribbon.

Step 2: Select Macros.

Step 3: Select View Macros.

Step 4: In the Macro dialog box, select the name of the macro that you want to change. We will use FillMonths for this example.

Step 5: Select Edit.

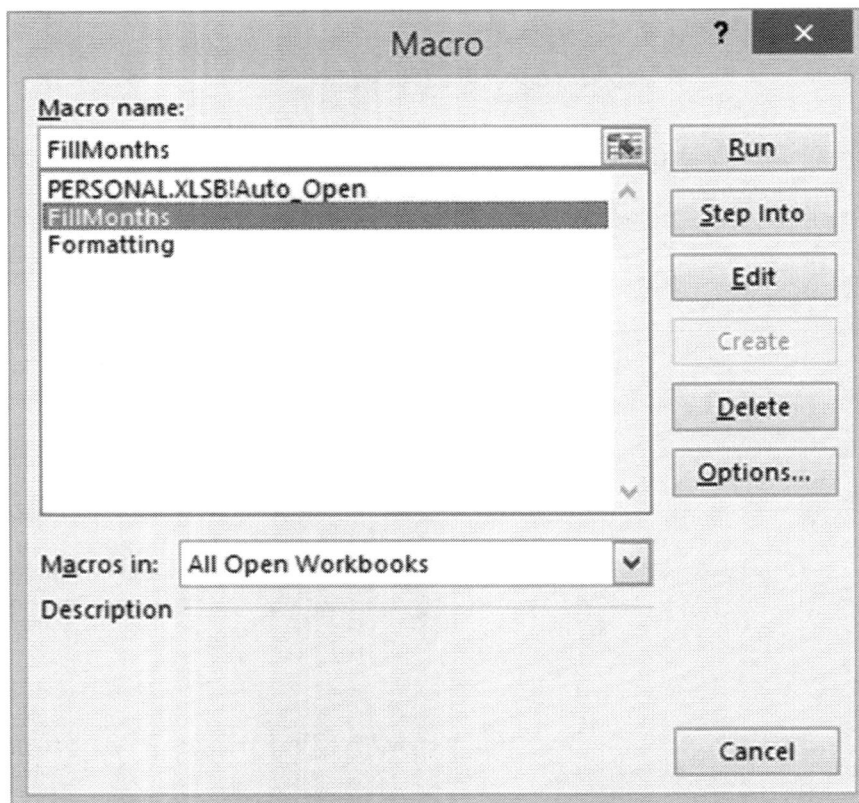

The Visual Basic Editor opens with your macro. The macro is a sub-routine in the programming. We will make a copy of our FillMonths() macro, or subroutine, and change both new macros slightly.

```
(General)                                                          ▼  FillMonths
Sub FillMonths()
'
' FillMonths Macro
'
'
    ActiveCell.Select
    ActiveCell.FormulaR1C1 = "Jan"
    ActiveCell.Select
    Selection.AutoFill Destination:=ActiveCell.Range("A1:L1"), Type:= _
        xlFillSeries
    ActiveCell.Range("A1:L1").Select
End Sub
```

First, we will make a copy of the macro.

Step 1: Select everything from Sub FillMonths() to End Sub.

Step 2: Select Copy. (CTRL + C)

Step 3: Place your cursor below End Sub.

Step 4: Paste by pressing CTRL + V.

```
(General)                                                          ▼  FillMonths
Sub FillMonths()
'
' FillMonths Macro
'
'
    ActiveCell.Select
    ActiveCell.FormulaR1C1 = "Jan"
    ActiveCell.Select
    Selection.AutoFill Destination:=ActiveCell.Range("A1:L1"), Type:= _
        xlFillSeries
    ActiveCell.Range("A1:L1").Select
End Sub

Sub FillMonths()
'
' FillMonths Macro
'
'
    ActiveCell.Select
    ActiveCell.FormulaR1C1 = "Jan"
    ActiveCell.Select
    Selection.AutoFill Destination:=ActiveCell.Range("A1:L1"), Type:= _
        xlFillSeries
    ActiveCell.Range("A1:L1").Select
End Sub
```

Now we are going to change the name of the first macro.

Step 1: At the top of the first subroutine, enter an R next to FillMonths, to represent, fill months row. Remember that macro names cannot contain spaces.

Now we are going to change the name of the second macro. Look for the line that divides the subroutines.

Step 2: Enter a C next to FillMonths, to represent fill months column.

Now we are going to change the second macro so that it fills down instead of across and it uses the number representation of the months instead of the names.

Step 3: In the following line, change "Jan " to 1

ActiveCell.FormulaR1C1 = "Jan"

The result should be

ActiveCell.FormulaR1C1 = 1

Step 4: Now change the range so that it is a column. You will need to change "A1:L1" to "A1:A12" in two places.

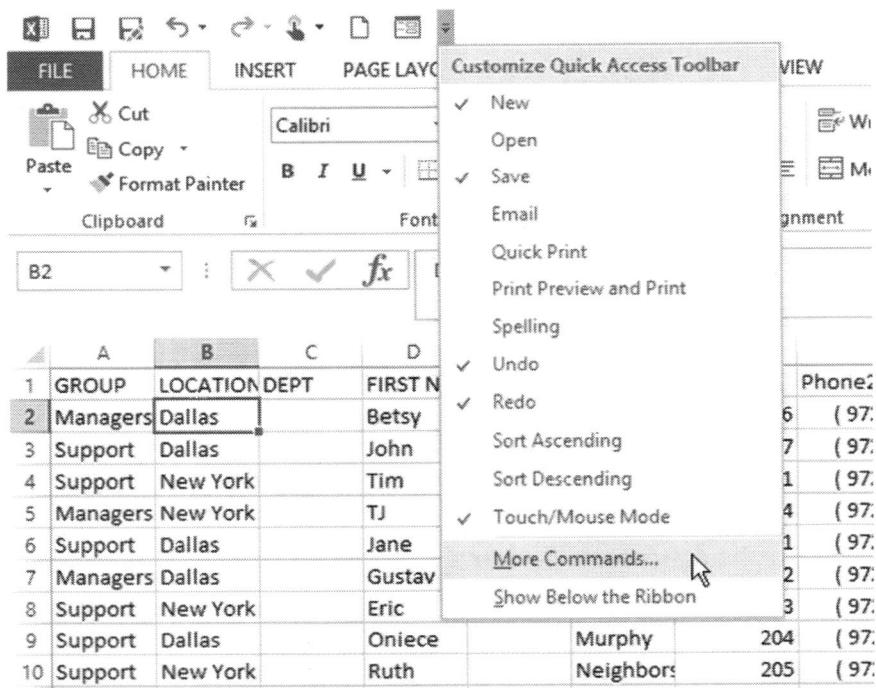

Now when you return to the Macros dialog box, you see that there are two separate macros with the names we assigned. Practice running them to see how they work.

This chapter takes a closer look at forms. Really, all Excel spreadsheets are a type of form where you can enter data. However, we will look at data forms and form controls little more closely as other ways to collect information with more flexibility. You will also learn how to create and use a data form, including adding a new row of data, finding information by navigating or by entering search criteria, and changing or deleting a row of data. Next, we will look at some specific examples of using different types of form controls. This chapter covers the list box control, the combo box control, the spin button control, and the scroll bar control.

About Excel Forms, Form Controls, and Active X Controls
Data Forms

Data forms are a convenient way to display one complete row of information in a range or table without having to scroll horizontally. Data forms work best when you have column headings that can work as your labels and you do not need sophisticated or custom form features.

Excel can automatically generate a built-in data form for a range or table in your worksheet. The form is displayed as a dialog box, in which each label has an adjacent blank text box. There is a maximum of 32 columns you can use for one data form.

Later in the chapter, you will learn how to create a data form, update rows, and delete rows. You will also learn how to navigate using the data form and how to find a row. Data forms can display a formula result as one of the items, but formulas cannot be changed using the data form.

Worksheets with Form and ActiveX Controls

Worksheets already contain some controls as part of the basic functionality of Excel. For example, you can create labels in cells and format them. You can use comments, hyperlinks, background images, data validation and conditional formatting, as well as other features, like controls.

However, you can also add several other controls and customize their properties to fine-tune your worksheet. For example, you can use a list box control to make it easier to select from a list of items.

Controls are placed on the drawing canvas of the worksheet, which means that the controls and objects are independent of row and column boundaries. However, you can also set controls to move and resize with a cell.

We will take about form controls and list controls. You can also use objects drawn with Drawing tools as controls, such as AutoShapes, WordArt, SmartArt and text boxes.

Form Controls

Here is a summary of the form controls in Excel 2016:

- Label – identifies the purpose of a cell or text box, or displays descriptive information like titles, captions, pictures or brief instructions
- Group box – groups related controls into one visual unit in a rectangle with a label (optional)
- Button – runs a macro that performs an action
- Check box – turns a value on or off, and is independent of other checkboxes that may appear in the same group
- Option (radio) button – allows a single choice within a set of mutually exclusive choices usually within a group box
- List box – displays a list of one or more items available as a choice and can be controlled as a single-selection list box, a multiple selection list box (must be adjacent choices) or an extended selection list box (allows multiple noncontiguous choices)
- Combo box – combines a text box with a list box to create a drop-down list box, which is like a list box, but more compact and requires the user to click the down arrow
- Scroll bar – allows the user to scroll through a range of values either by clicking the arrows or dragging the scroll bar

- Spin button – increases or decreases a value by clicking the up or down arrow

ActiveX Controls

ActiveX controls are much like form controls, except they allow more flexible design requirements, including appearance, behavior, fonts, and other characteristics. Many can be used with or without VBA code, although some can only be used with VBA UserForms. You cannot use ActiveX controls on chart sheets or XLM macro sheets. Here is a summary of the ActiveX controls in Excel 2016:

- Check box – turns a value on or off, and is independent of other checkboxes that may appear in the same group
- Text box – enables you to view type or edit text or data bound to a cell
- Button – runs a macro that performs an action. You cannot assign a macro to run directly from an ActiveX button the same way you can from a form control.
- Option (radio) button – allows a single choice within a set of mutually exclusive choices usually within a group box
- List box – displays a list of one or more items available as a choice and can be controlled as a single-selection list box, a multiple selection list box (must be adjacent choices) or an extended selection list box (allows multiple noncontiguous choices)
- Combo box – combines a text box with a list box to create a drop-down list box, which is like a list box, but more compact and requires the user to click the down arrow
- Toggle button – indicates a state (like yes or no or on or off) that alternates between enabled and disabled
- Spin button – increases or decreases a value by clicking the up or down arrow
- Scroll bar – allows the user to scroll through a range of values either by clicking the arrows or dragging the scroll bar
- Label – identifies the purpose of a cell or text box, or displays descriptive information like titles, captions, pictures or brief instructions

- Image – embeds a picture file (such as bitmap, JPEG, or GIF)

- More controls – allows you to choose from additional controls

Using a Data Form

You will need to add the Form button to the Ribbon or the Quick Access Toolbar, use the following procedure to add it to the Quick Access Toolbar.

Step 1: Select the arrow by the Quick Access Toolbar.

Step 2: Select More Commands.

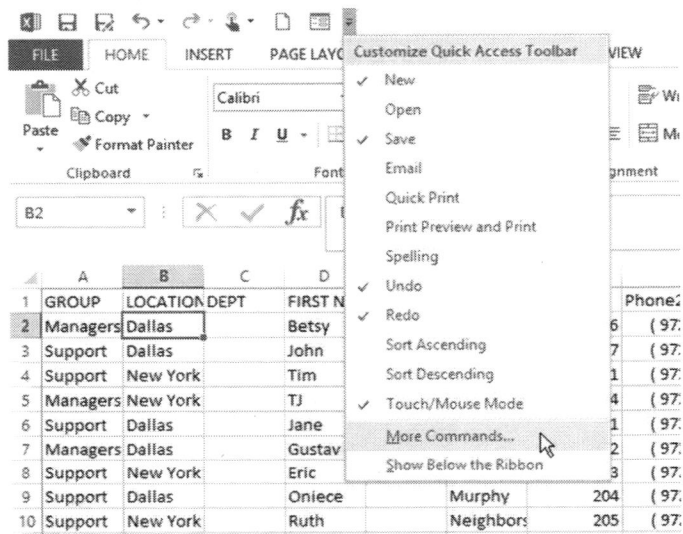

Step 3: In the Excel Options dialog box, select All Commands from the Choose Commands from list.

Step 4: Select Form from the list on the left.

Step 5: Select Add.

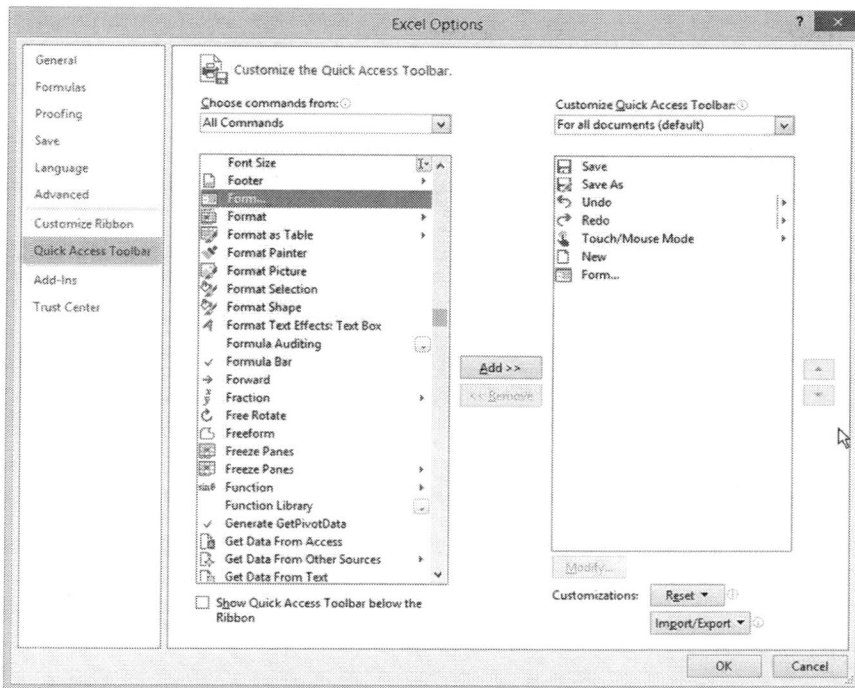

Step 6: Select OK to close the dialog box.

Now you can create your form, use the following procedure.

Step 7: With your cursor anywhere in the data on the worksheet, select the Form tool from the Quick Access Toolbar.

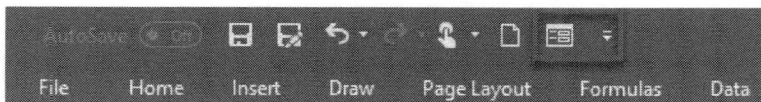

Your form is automatically created.

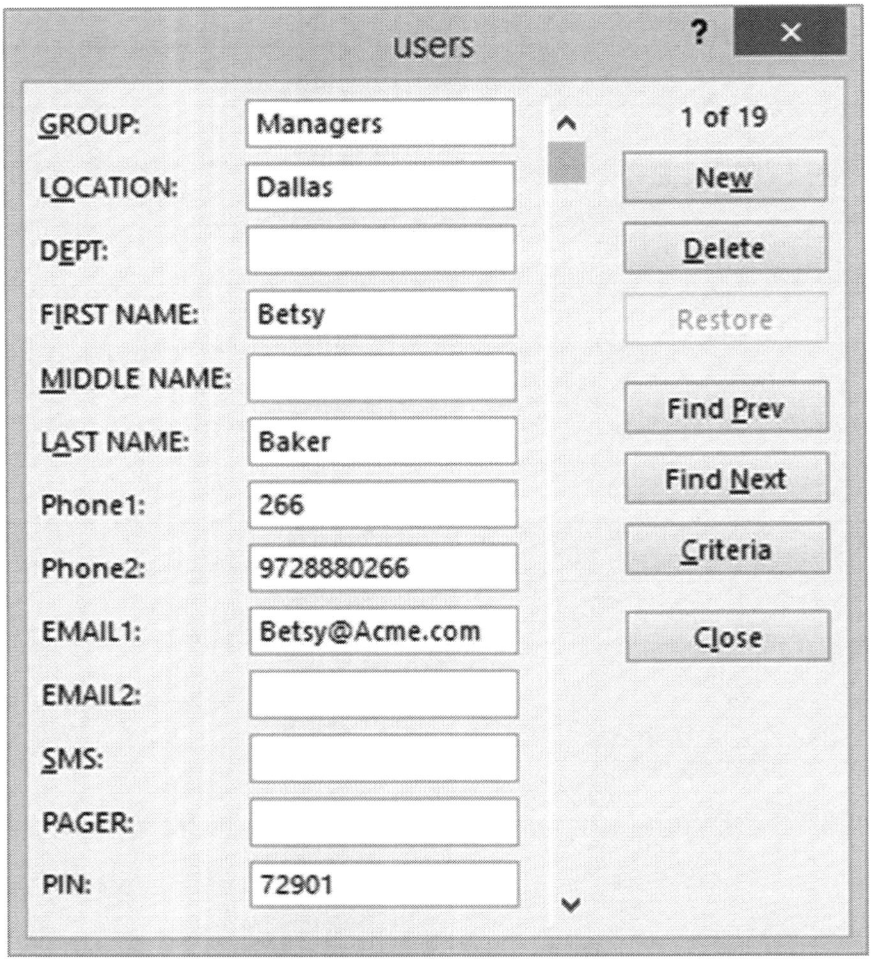

users		?	X

GROUP: `Managers` ^ 1 of 19

LOCATION: `Dallas`   **New**

DEPT: ` `   **Delete**

FIRST NAME: `Betsy`   Restore

MIDDLE NAME: ` `

LAST NAME: `Baker`   **Find Prev**

Phone1: `266`   **Find Next**

Phone2: `9728880266`   **Criteria**

EMAIL1: `Betsy@Acme.com`   **Close**

EMAIL2: ` `

SMS: ` `

PAGER: ` `

PIN: `72901` v

To add a new row of data, use the following procedure.

Step 1: In the data form, select New.

Step 2: Enter the information into each text field. You can press the TAB key to go to the next field. Press Enter to complete the record and go to another new record. Excel extended the worksheet behind the form down for each new record you create.

To find a row by navigating. Use the following controls:

- The scroll bar allows you to move through one row at a time. Select the up or down arrows to move through the data.
- The scroll bar also allows you to move through the data 10 rows at a time. Select the scroll bar in the area in between the arrows.
- Find Prev allows you to move to the previous row in the data.
- Find Next allows you to move to the next row in the data.

To find a row by searching, use the following procedure.

Step 1: Select Criteria.

Excel displays a blank form.

Step 2: Enter information in one or more fields to indicate your criteria. You can select Clear to start over. You can also use the following wildcards:

- ? to replace a single character (sm?th finds "smith" and "smyth")
- * to find any number of characters (*east finds "northeast" and "southeast")
- ~ followed by wildcard character finds a question mark, an asterisk or a tilde Step 2: You can now scroll (Find Prev or Find Next) through any matching records.

Step 3: Select Form to return to the form.

To change the data in a row, use the following procedure.

Step 1: If you have not yet entered the data by pressing Enter, you can select Restore to return the data to what is stored on the worksheet.

Step 2: Otherwise, first find the row you want to change. Then simply type over the old data.

Step 3: Press Enter to update the row.

To delete a row, use the following procedure.

Step 1: Find the row that you want to delete.

Step 2: Select Delete.

Step 3: In the confirmation message, select OK. Note that you cannot undo a row deletion after you have confirmed it.

Using a List Box Control

To insert a list box form control, use the following procedure.

Step 1: Make sure the Developer tab is showing. If not, go to Options, Customize Ribbon and check the Developer checkbox.

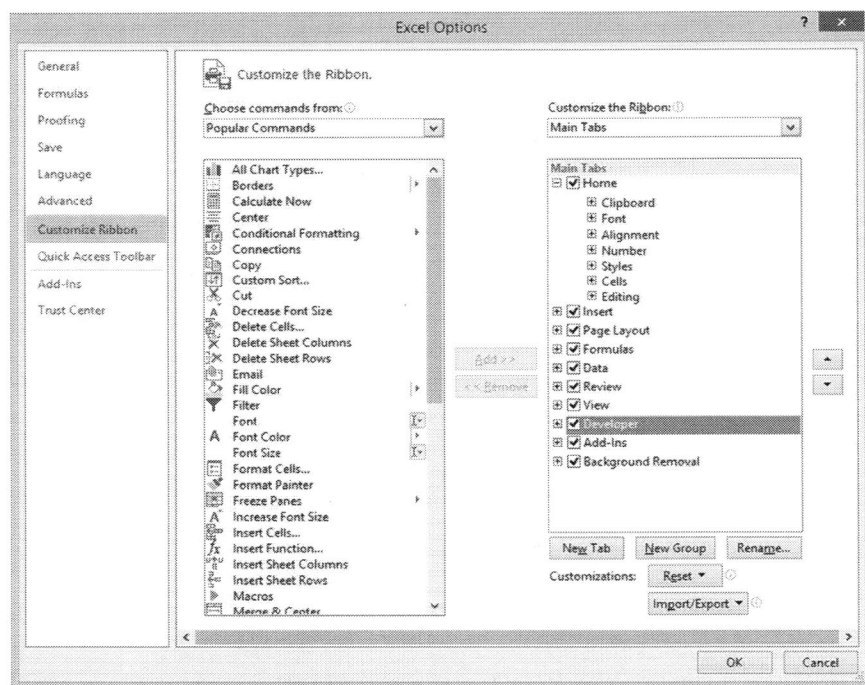

Step 2: Select the Developer tab on the Ribbon.

Step 3: Select Insert. Select List Box Form from the drop-down list.

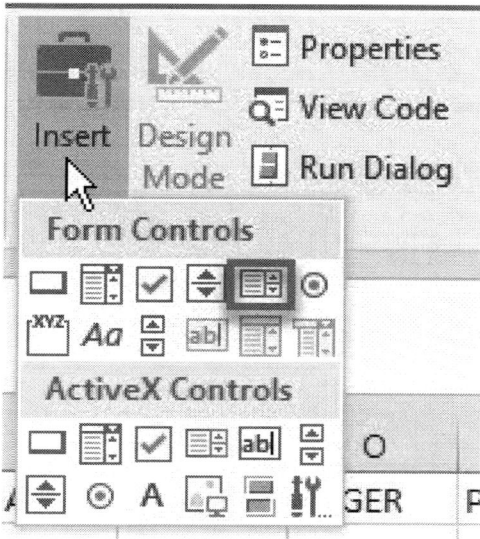

Step 4: Click and drag the mouse to draw the list box. In this example, start at B2 and drag down to E10.

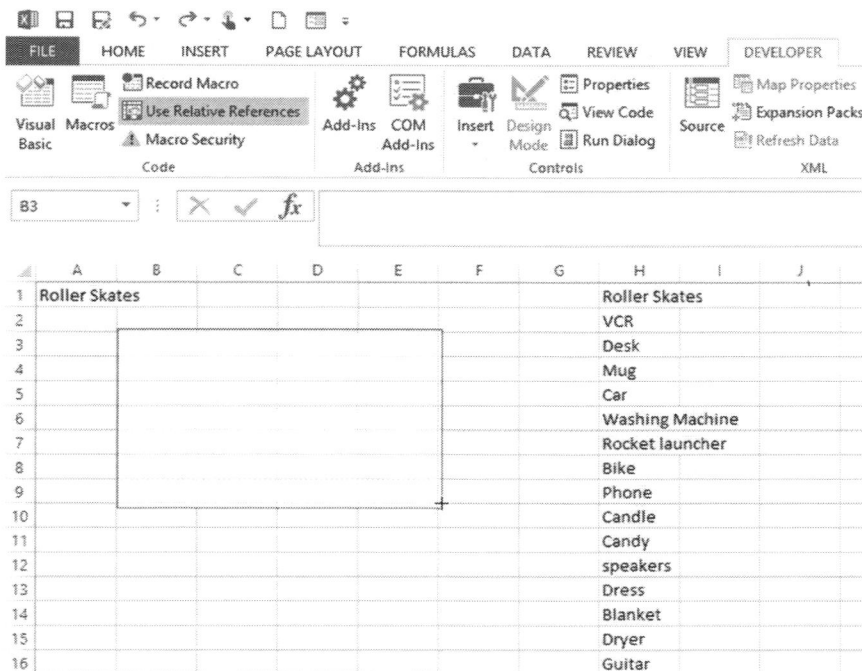

Now we need to format the list box. With the list box still selected, select Properties from the Developer tab on the Ribbon.

The Control tab of the Format Control dialog box allows you to indicate an input range and a cell link.

Step 5: In the Input range field, enter or select cells H1:H20.

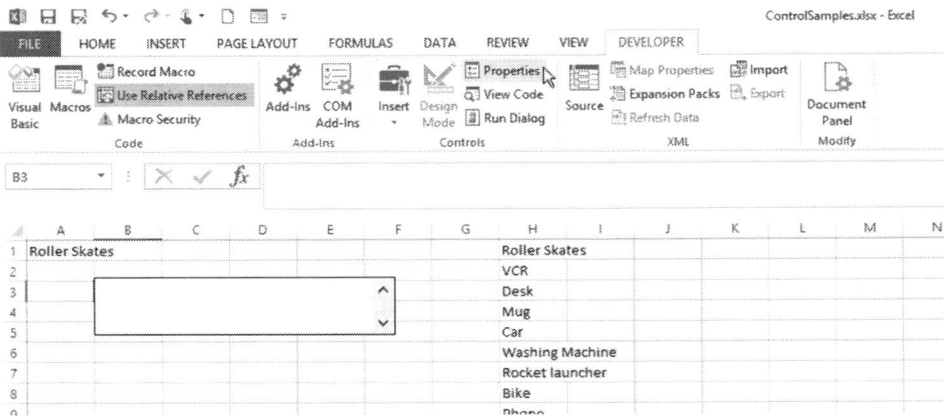

Step 6: The Cell link field will put a number value in the linked cell. We will do this so our INDEX formula (which references cell G1) in cell A1 will return the list box choice. Enter or select G1 for the Cell link field.

Step 7: Make sure the Single option is selected.

Step 8: Select OK.

Step 9: Now click anywhere besides the list box to deselect it. When you select an item in the list, the INDEX formula in cell A1 uses the selection to display the item's name.

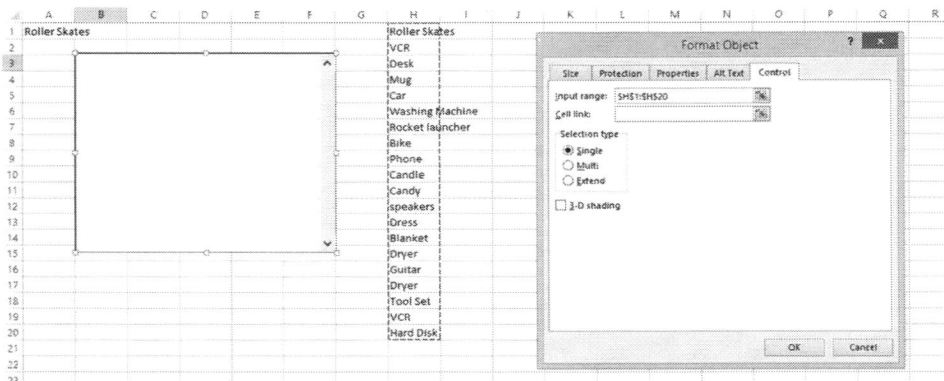

Using a Combo Box Control

To insert a combo box form control, use the following procedure.

Step 1: Select the Developer tab.

Step 2: Select Insert. Select Combo Box Form from the drop-down list.

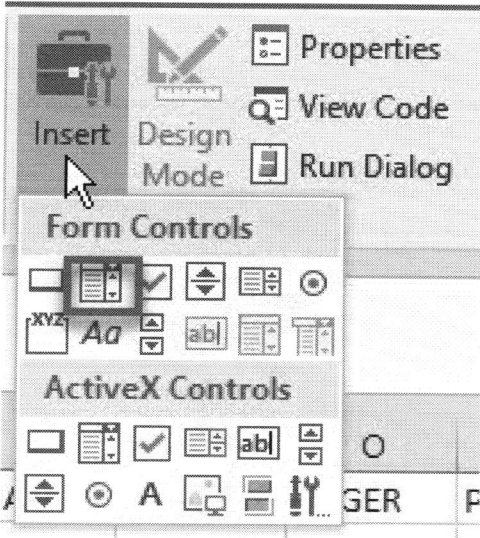

Step 3: Click and drag the mouse to draw the combo box.

Step 4: Right-click the combo box, and select Format Control from the drop-down list.

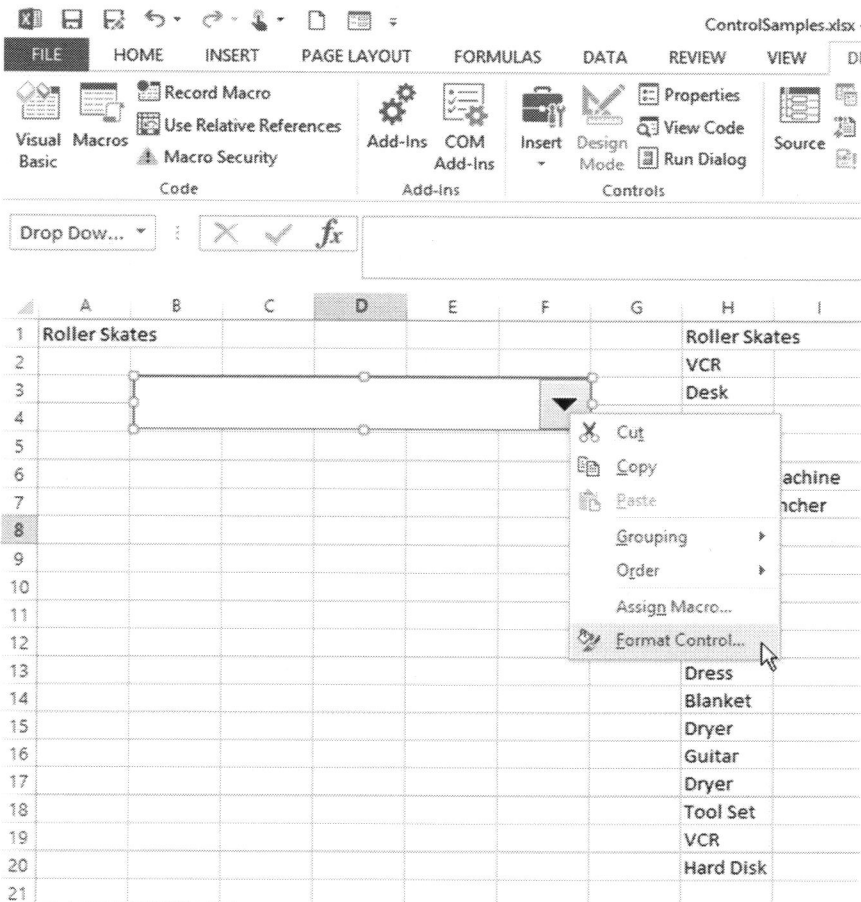

Step 5: In the Input range field, enter or select cells H1:H20.

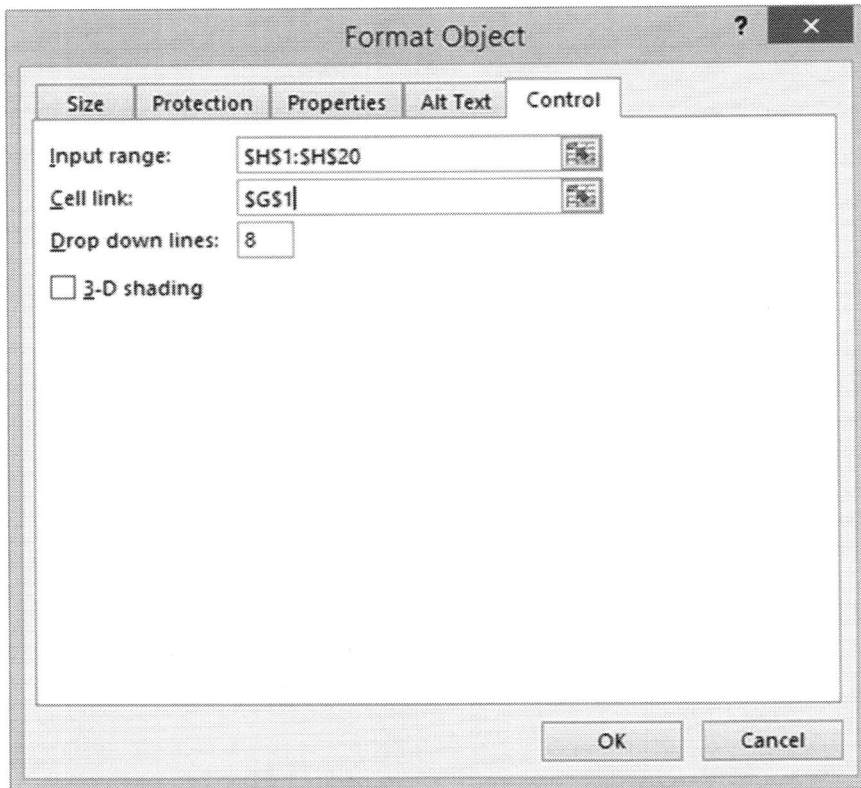

Step 6: The Cell link field will put a number value in the linked cell. We will do this so our INDEX formula (which references cell G1) in cell A1 will return the list box choice. Enter or select G1 for the Cell link field.

Step 7: The Drop-down lines indicates the number of lines that show in the list at one time.

Step 8: Select OK.

Step 9: Now click anywhere besides the combo box to deselect it. When you select the arrow, and choose an item from the list, the INDEX formula in cell A1 uses the selection to display the item's name.

Using a Spin Button Control

To insert a list spin button control, use the following procedure.

Step 1: Select the Developer tab.

Step 2: Select Insert. Select Spin Button from the drop-down list.

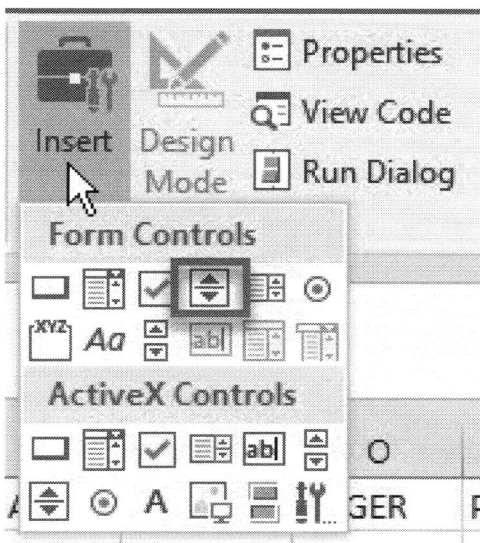

Step 3: Click and drag the mouse to draw the spin button control.

Step 4: Right-click the spin button control, and select Format Control from the drop-down list.

Step 5: On the Control tab of the Format Control dialog box, enter 1 in the Current Value field.

Step 6: Also enter 1 as the Minimum Value to restrict the top of the spin button to the first item in our list.

Step 7: Enter 20 as the Maximum value to specify the maximum number of entries in the list.

Step 8: Enter 1 as the Incremental Change.

Step 9: To put a number value in cell G1 for our INDEX formula in this example, enter G1 in the Cell link field.

Step 10: Select OK.

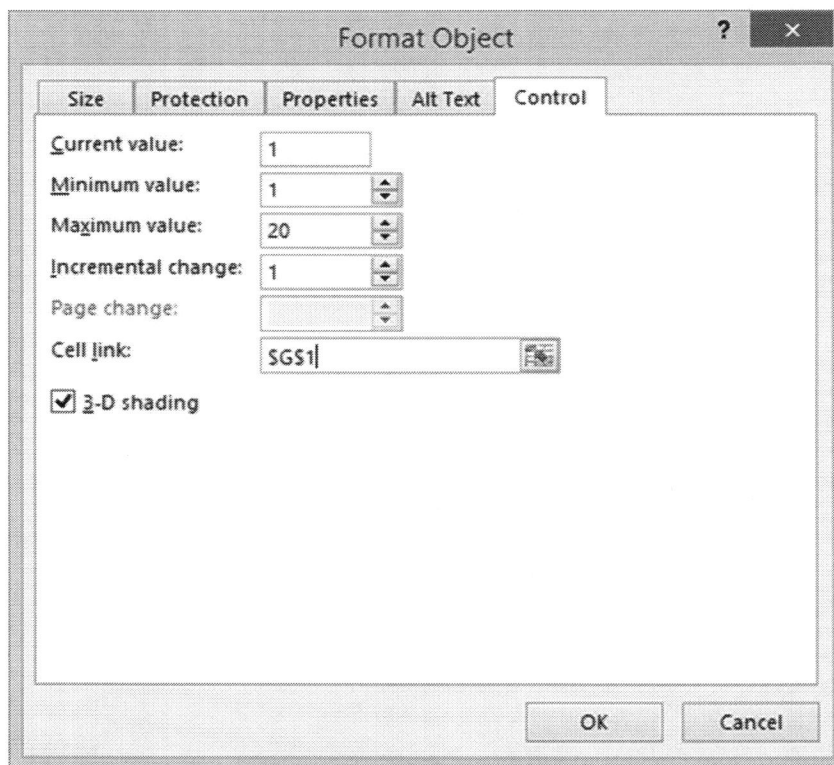

Step 11: Now click anywhere besides the spin button to deselect it. When you click the up or down controls, cell G1 is updated. The INDEX formula in cell A1 uses the value in G1 to display the item's name.

Using a Scroll Bar Control

To insert a list scroll bar control, use the following procedure.

Step 1: Select the Developer tab.

Step 2: Select Insert. Select Scroll Bar from the drop-down list.

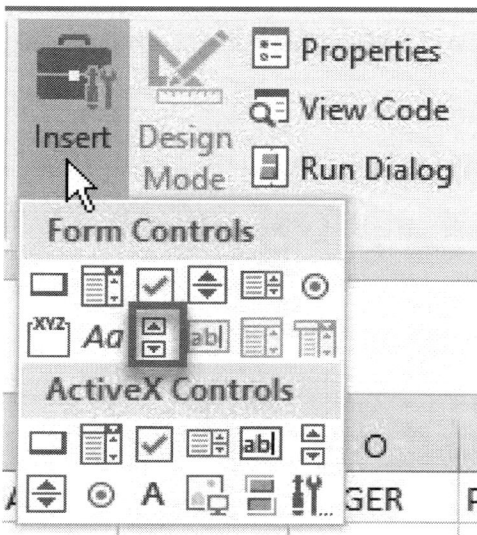

Step 3: Click and drag the mouse to draw the scroll bar control.

Step 4: Right-click the scroll bar control, and select Format Control from the drop-down list.

Step 5: On the Control tab of the Format Control dialog box, enter 1 in the Current Value field.

Step 6: Also enter 1 as the Minimum Value to restrict the top of the scroll bar to the first item in our list.

Step 7: Enter 20 as the Maximum value to specify the maximum number of entries in the list.

Step 8: Enter 1 as the Incremental Change.

Step 9: In the Page Change field, enter 5 to control how much the current value increments if you click inside the scroll bar instead of on the arrows.

Step 10: To put a number value in cell G1 for our INDEX formula in this example, enter G1 in the Cell link field.

Step 11: Select OK.

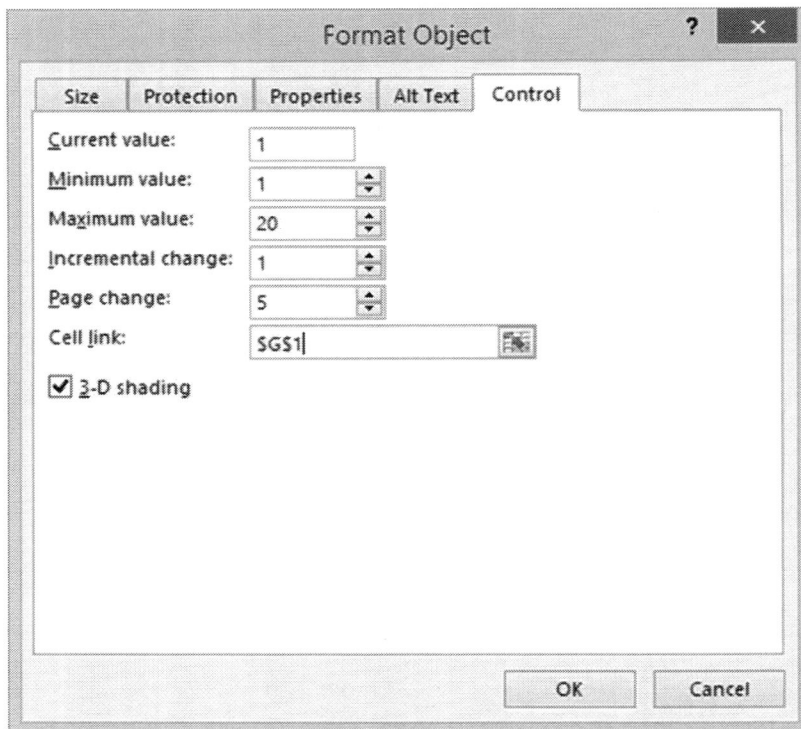

Step 12: Now click anywhere besides the scroll bar to deselect it. When you click the up or down arrows or anywhere in the scroll bar, cell G1 is updated. The INDEX formula in cell A1 uses the value in G1 to display the item's name.

Chapter 27 – Applying Advanced Chart Features

In this chapter, you will learn about some advanced chart features. First, we will look at the different types of trend lines to help you analyze your data. You will also learn how to add a trend line. Then, we will look at how to plot one of your data series on a secondary axis. Finally, you will learn how to save a chart you have formatted to your liking as a chart template to be available for use when creating other new charts.

About Trend Lines

You can add one of six different trend or regression types, depending on the type of data that you have.

Note that a trend line is most accurate when its R-squared value is at or near 1. You can display the R-squared value that Excel calculates on your chart.

The following types of trend lines are available.

Linear trend line – this trend line is a best-fit straight line used with simple linear sets.

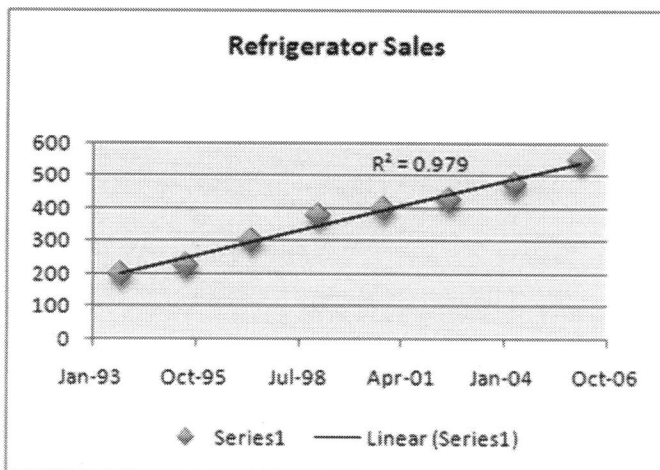

Logarithmic trend line – this trend line is a best-fit curved line used when the rate of change in the data increases or decreases quickly and then levels out. It can use either negative or positive values.

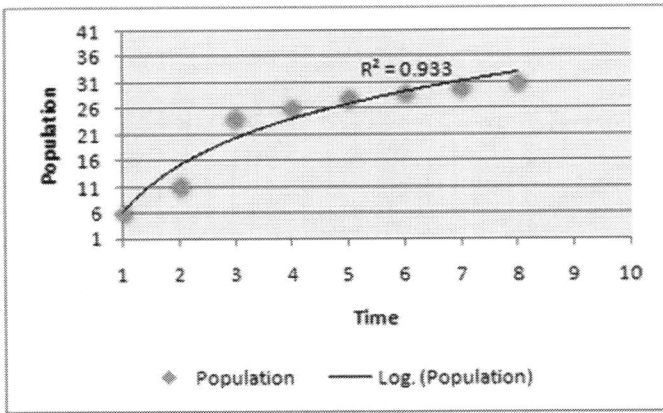

Polynomial trend line – this trend line is a curved line that is used when data fluctuates, such as to analyze gains and losses over a large data set. The polynomial order can be determined by the number of fluctuations in the data or by how many hills and valleys appear in the curve. For example, an Order 2 polynomial trend line has one hill or valley; order 3 has 2 hills or valley, and so on.

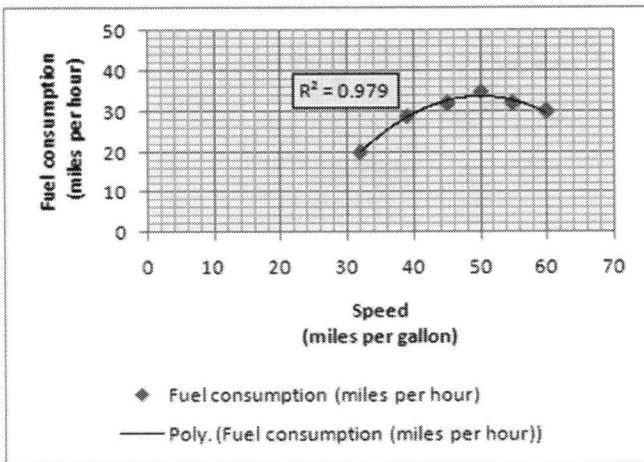

Power trend line – this trend line is a curved line that is used with data sets that compare measurements that increase at a specific rate. Data sets for this type of trend line cannot include 0 or negative values.

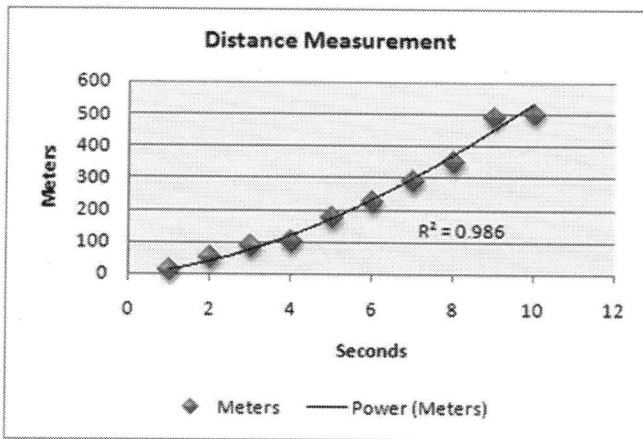

Exponential trend line – this trend line is a curved line that is used with data sets that rise or fall at a constantly increasing rate. Data sets for this type of trend line cannot include 0 or negative values.

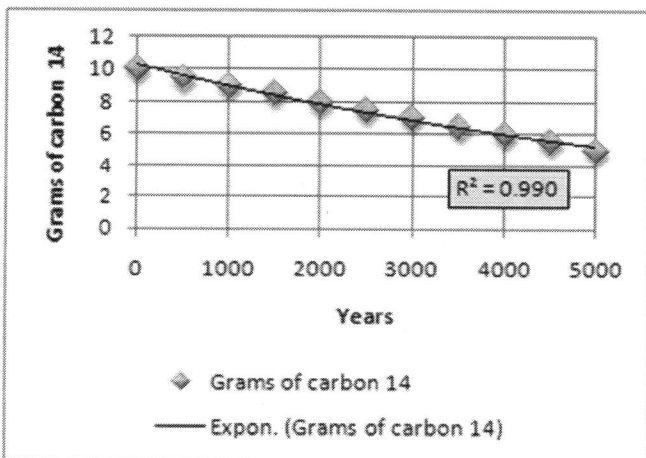

Moving Average Trend Lines – this trend line smooths fluctuations in data to show a pattern or trend more clearly. It uses a specific number of data points in an average to display the points in the line, such as 2 periods to create the first point in the trend line.

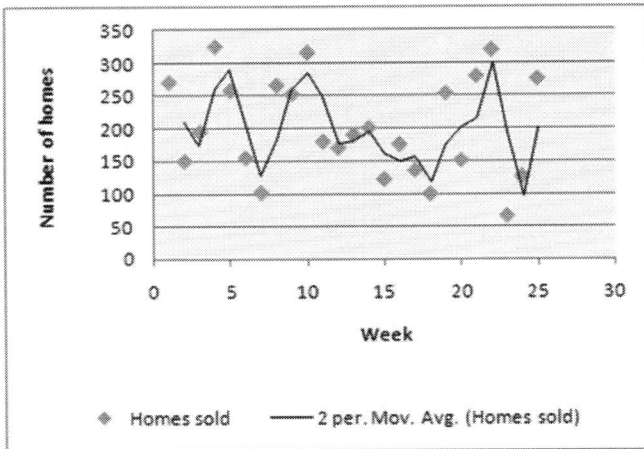

Add a Trend Line

To add a Trend Line to a chart, use the following procedure.

Step 1: Select the chart in your worksheet.

Step 2: Select the data series that you want to plot. A colored border appears around that area in the data.

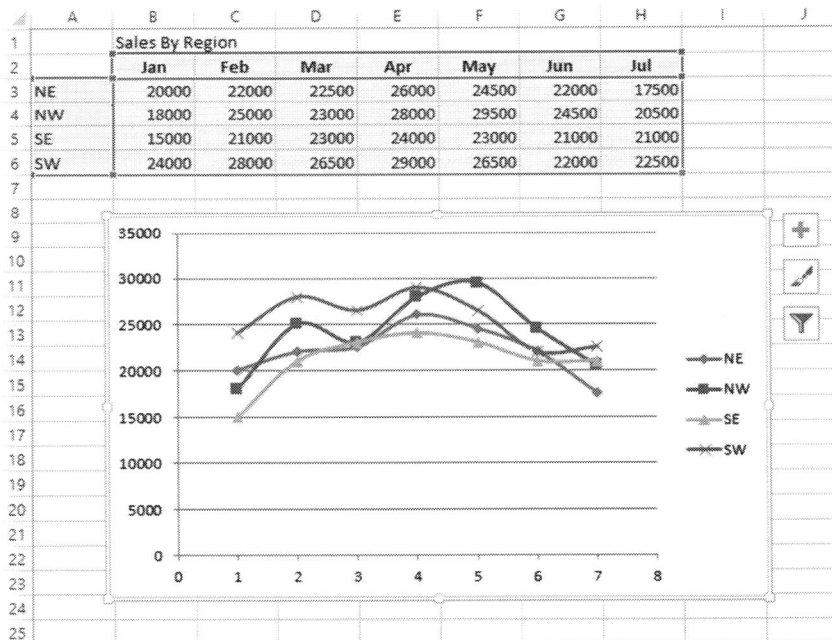

Step 3: Select the plus sign icon from the right side of the chart.

Step 4: Select the small arrow to the right of Trendlines.

Step 5: Select the type of trendline that you want to add from the list.

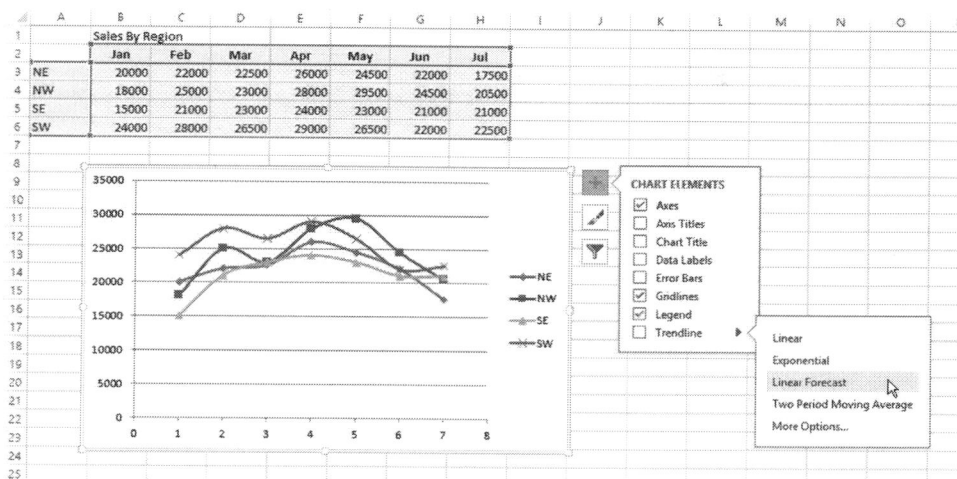

Step 6: If you did not select the specific data series and the chart has more than one, Excel displays a dialog box to help you choose the correct series.

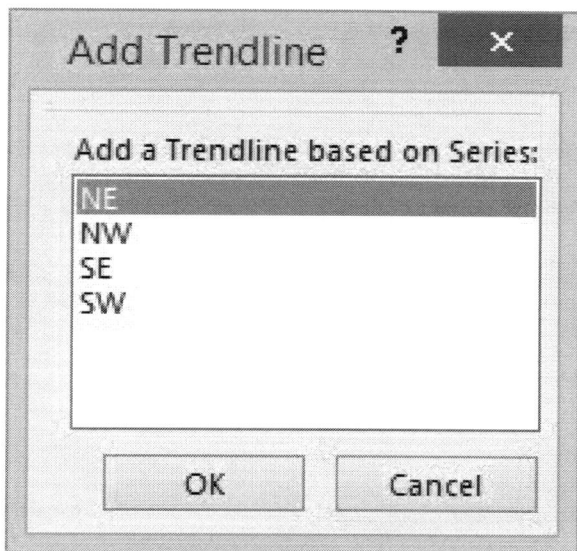

Excel adds the Trendline.

Note that if you select More Options from the Trendline choices, Excel will open the Format Trendline pane, where you have several additional options for formatting your trendline.

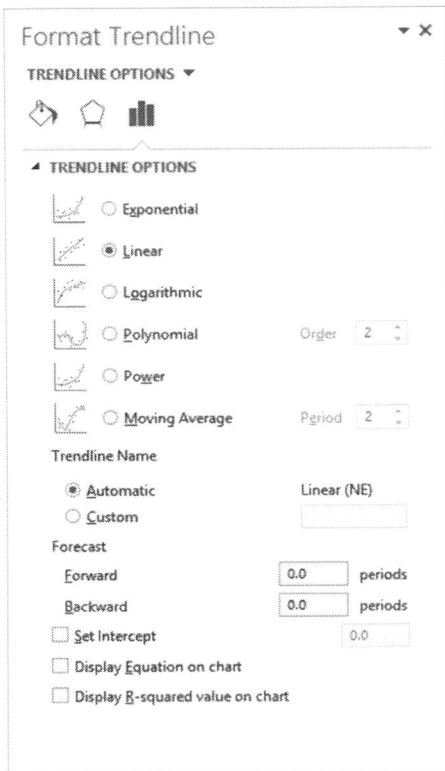

Using Secondary Axes

To add a secondary vertical axis, use the following procedure.

Step 1: Select the chart.

Step 2: Select the Chart Tools Format tab.

Step 3: Select the Chart Elements box arrow and select the data series that you want to plot along a secondary vertical axis.

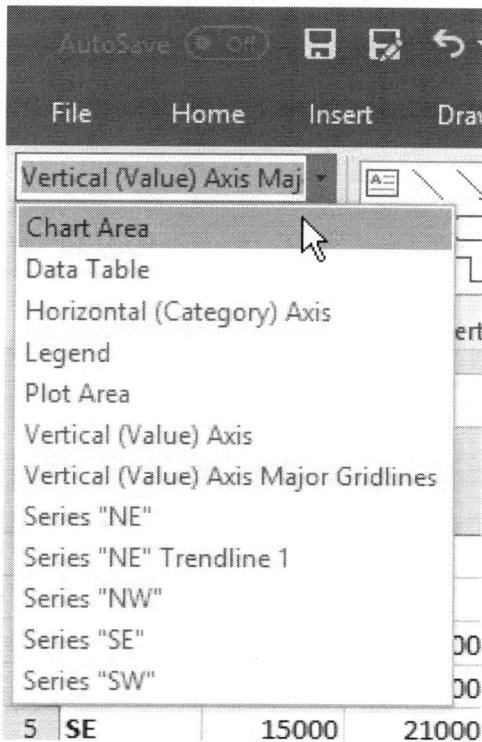

Step 4: Select Format Selection from the Chart Tools Format tab on the Ribbon.

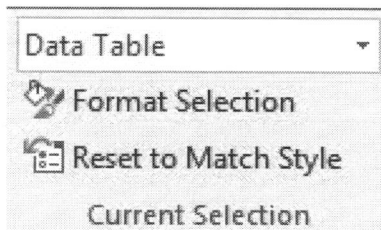

Step 5: Excel displays the Format Data Series pane. In the Format Data Series pane, select Secondary Axis under Plot Series On. The Format Data Series pane stays open so that you can format other aspects of the selected data series if desired.

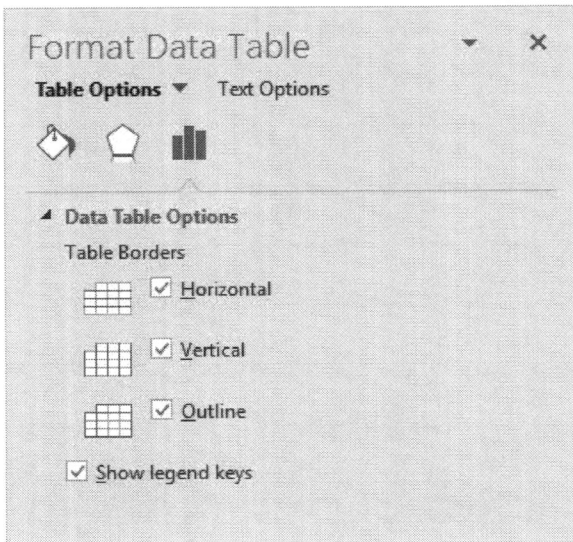

In the following illustration, the chart type for the secondary axis has also been changed. With the secondary axis still selected, just go to the Chart Tools Design tab and select Change Chart Type.

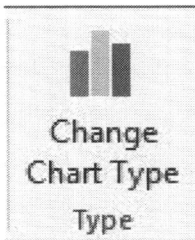

In the Change Chart Type dialog box, your chart is automatically shown as a Custom Combination chart. You can use the drop down lists next to each data series to select another chart option for that data series.

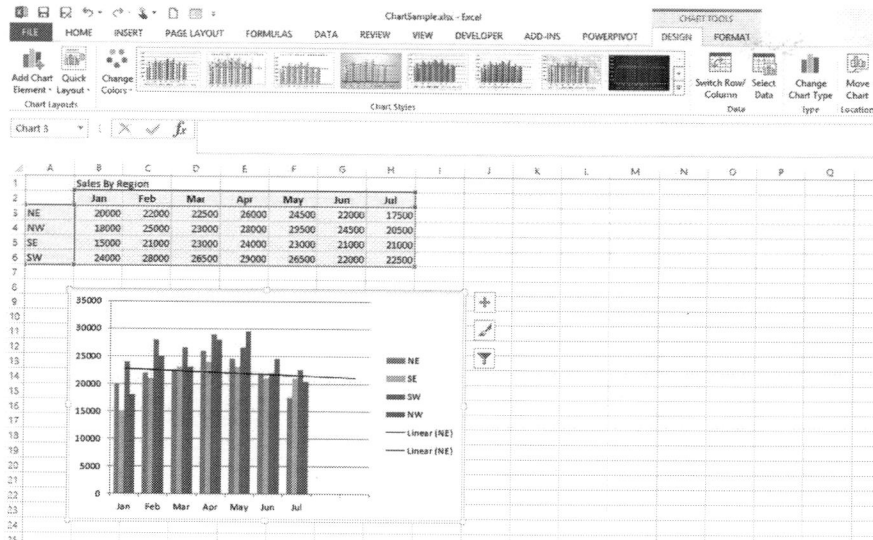

Using Chart Templates

To save a chart as a template, use the following procedure.

Step 1: Right-click the chart.

Step 2: Select Save As Template.

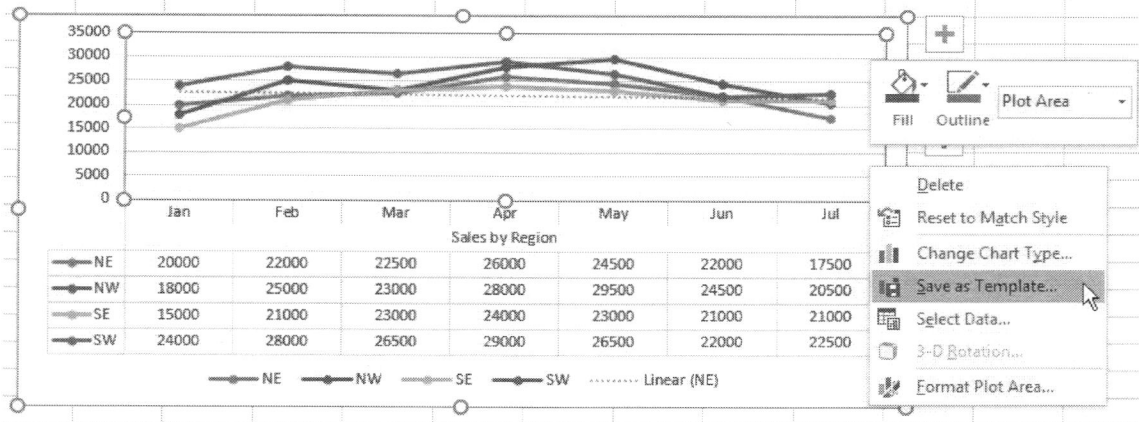

Step 3: In the Save Chart Template dialog box, enter a File name for the template.

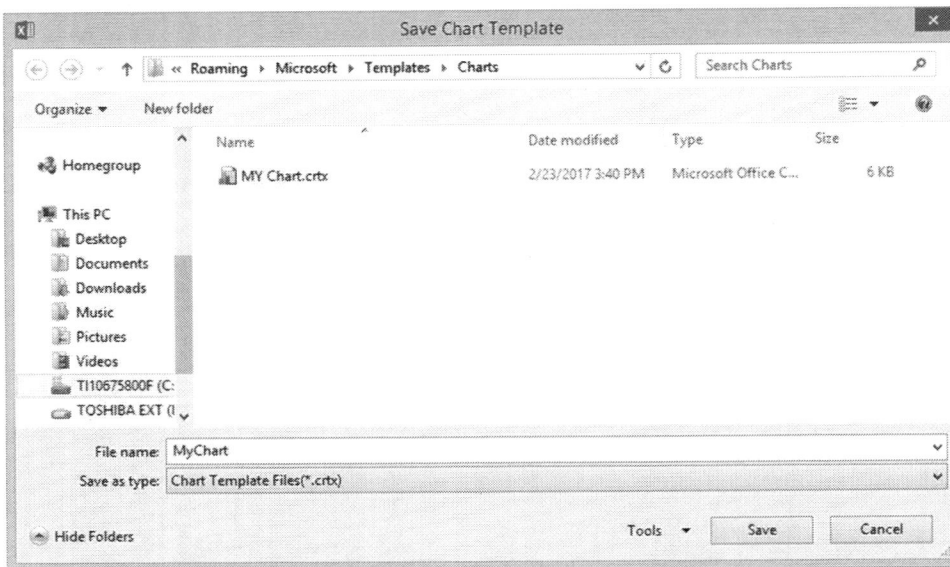

Step 4: Leave the file location as is, if you want the chart template to be available on the Insert tab when you select Charts or when you select Change Chart Type.

Step 5: Select Save.

To insert a chart based on their templates, use the following procedure.

Step 1: Select the data you want to use for your chart. You may want to copy the data from the Secondary Axis or Trendlines sheet in the sample file to Sheet 3 for this example.

Step 2: Select the Insert tab from the Ribbon.

Step 3: Select Recommended Charts.

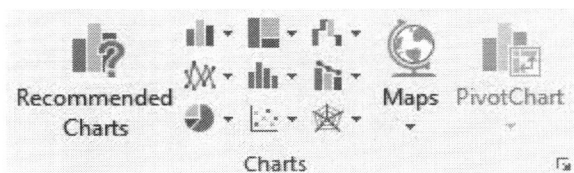

Step 4: Select the All Charts tab.

Step 5: Select the Templates option from the left.

Step 6: Select the template that you have previously saved.

Step 7: Select OK to insert the chart.

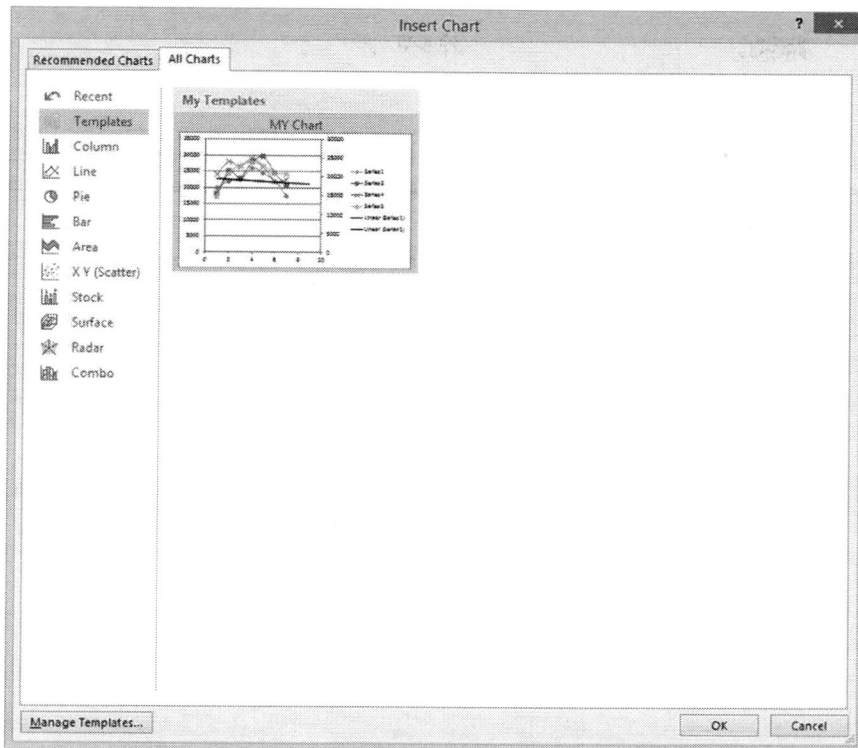

This chapter explains slicers and timelines. You will learn how to create a slicer in an existing table. You will also learn how to format a slicer. Finally, we will discuss how to disconnect or delete a slicer.

About Slicers and Timelines

Slicers were added in Excel2010 to filter PivotTable data. In Excel 2016, you can now use slicers to filter any table data. Slicers clearly indicate what data is shown in the table after you filter the data. They include buttons so that you can quickly filter data without having to use drop down lists to find the items you want to filter. Timelines are a type of slicer that is specific to date ranges.

A slicer typically displays the following elements:

- The slicer header – indicates the category of the items in the slicer.
- Unselected filtering button – indicates that the item is not included in the filter.
- Selected filtering button – indicates that the item is included in the filter.
- Clear Filter button – removes the filter by selecting all items in the slicer.
- A scroll bar – enables scrolling when there are more items than are currently visible in the slicer.
- Border moving and resizing controls – allow you to change the size and location of the slicer.

Using slicers

To filter your data, just select one or more of the buttons in the slicer.

You will likely that you will create more than once slicer to filter a data table or PivotTable report.

You can create a slicer that is associated with the current data table or PivotTable. You can also create a copy of a slicer.

Once you create a slicer, it appears on the worksheet alongside the table data, in a layered display if you have more than one. You can move or resize it as needed. Once created, a slicer can also be used with another table or PivotTable.

You can create slicers that work with the current data table or PivotTable or you can create a stand-alone slicer that can be associated with any other table later. Stand-alone slicers can be referenced by Online Analytical Processing (OLAP) Cube functions.

Creating a Slicer in an Existing Table

To create a slicer in an existing table, use the following procedure.

Step 1: Place your cursor anywhere in the table.

Step 2: Select the Insert tab from the Ribbon.

Step 3: Select Slicer.

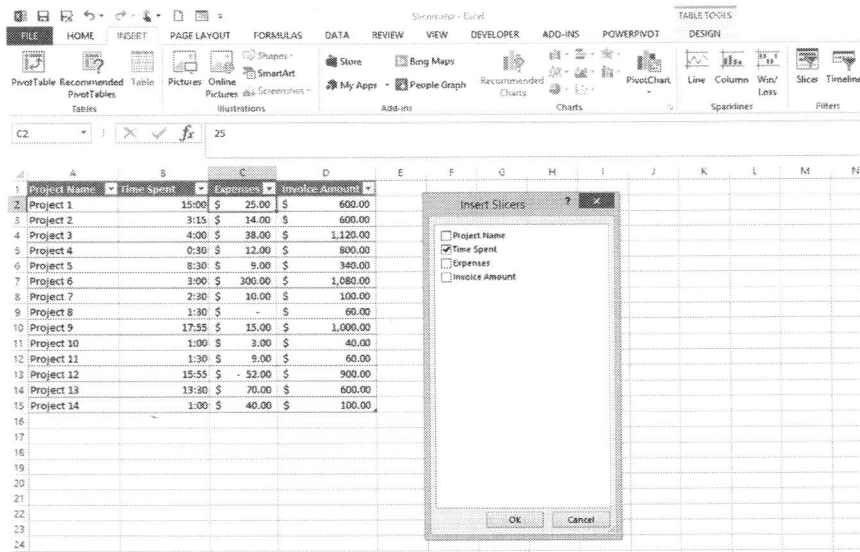

Step 4: In the Insert Slicers dialog box, select the check boxes of the fields from the table for which you want to create a slicer.

Step 5: Select OK.

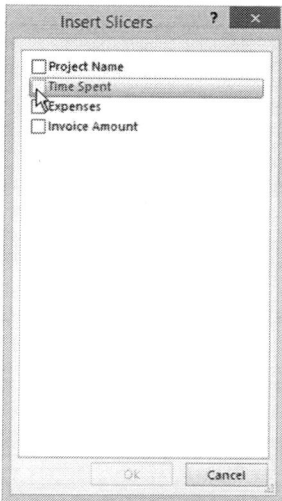

Step 6: To apply your filters, just select the buttons for the items you want to include. You can hold down the CTRL key while selecting to choose more than one button.

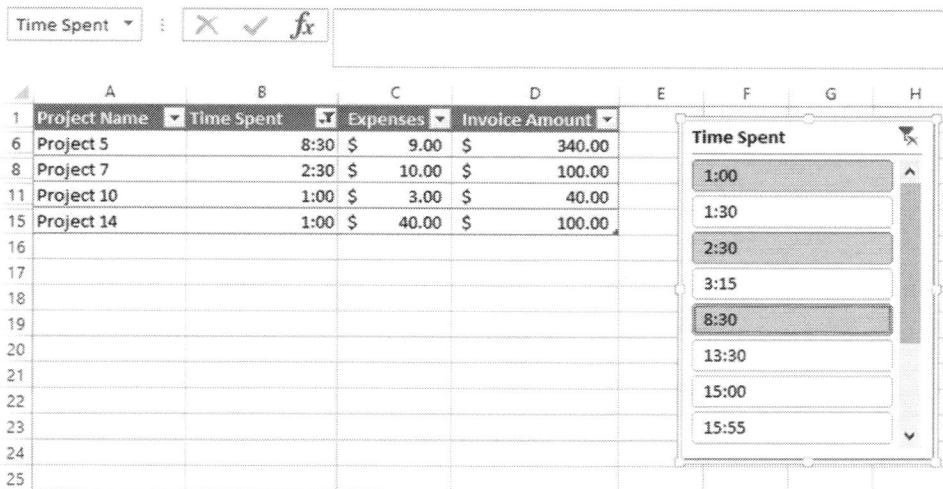

Formatting a Slicer

Note that when you select the slicer object, the Slicer Tools Options tab displays on the Ribbon, giving you additional tools to adjust your slicer settings, apply a new slicer style, or arrange and size the slicer object or the buttons.

To format a slicer, use the following procedure.

Step 1: Select the slicer that you want to format.

Step 2: Select the Slicer Tools Options tab from the Ribbon.

Step 3: You can select a new Style from the Slicer Styles area.

Step 4: Or, you can select the arrows next to Slicer Styles and select New Slicer Style.

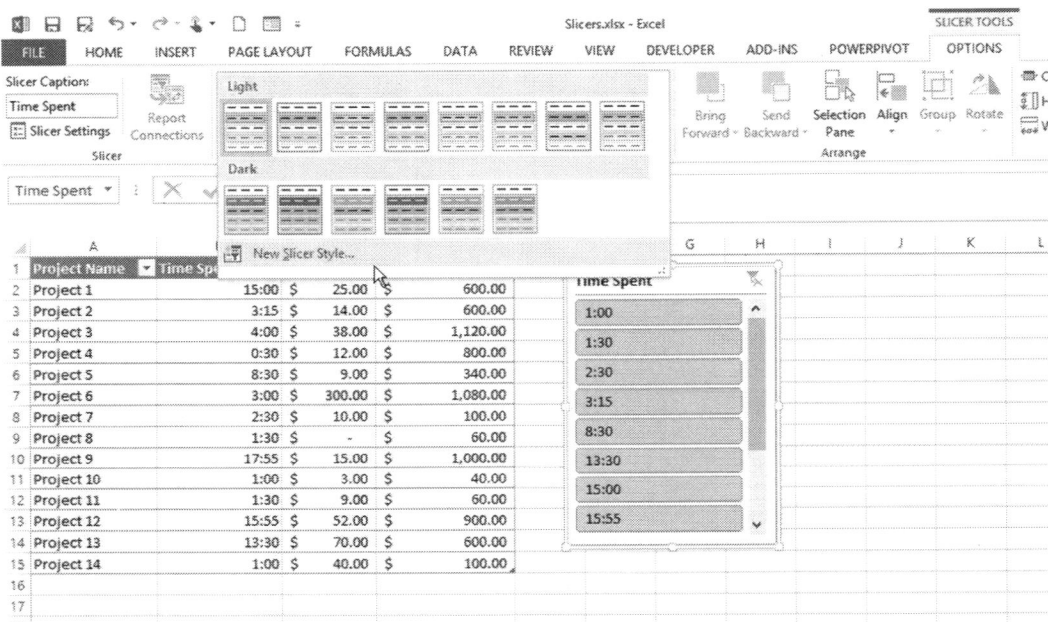

Step 5: In the New Slicer Style dialog box, enter a Name for the new style.

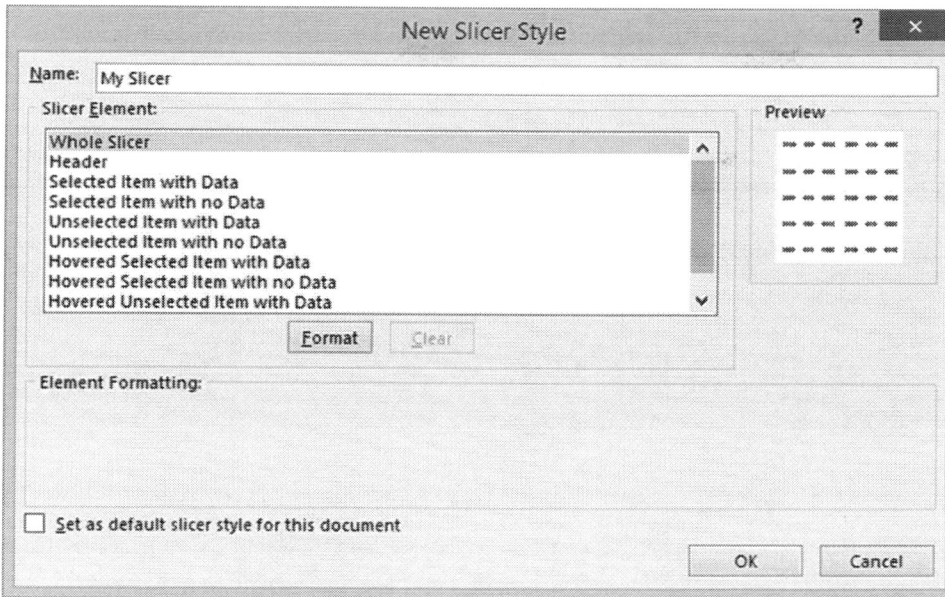

Step 6: Select the Slicer Element that you want to format from the list.

Step 7: Select Format. You can change the Font, Border, and Fill colors for each element.

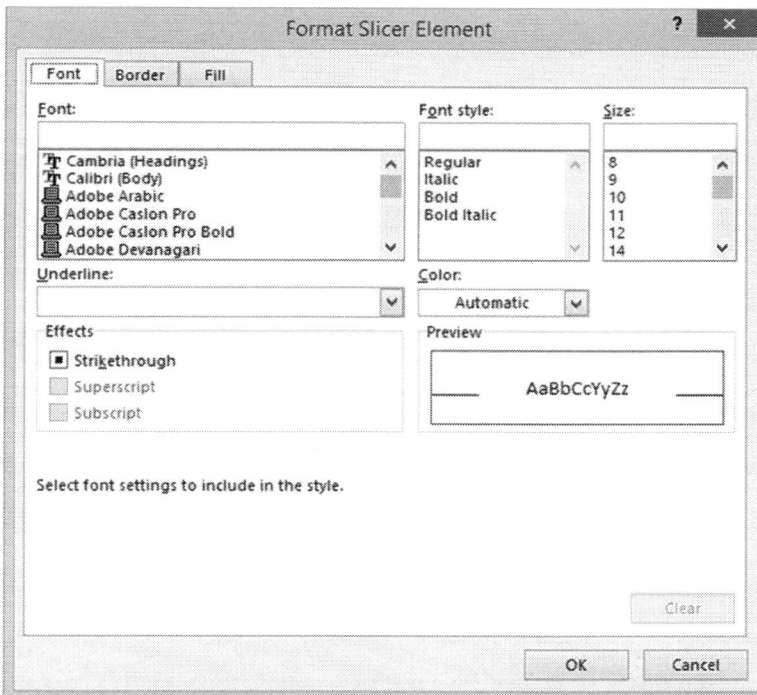

Step 8: Select OK when you have finished formatting the selected element.

Step 9: Select OK when you have finished formatting all the desired elements. You can check the Set as default slicer quick style for this document check box if desired.

Practice adjusting the size for the slicer and buttons as well.

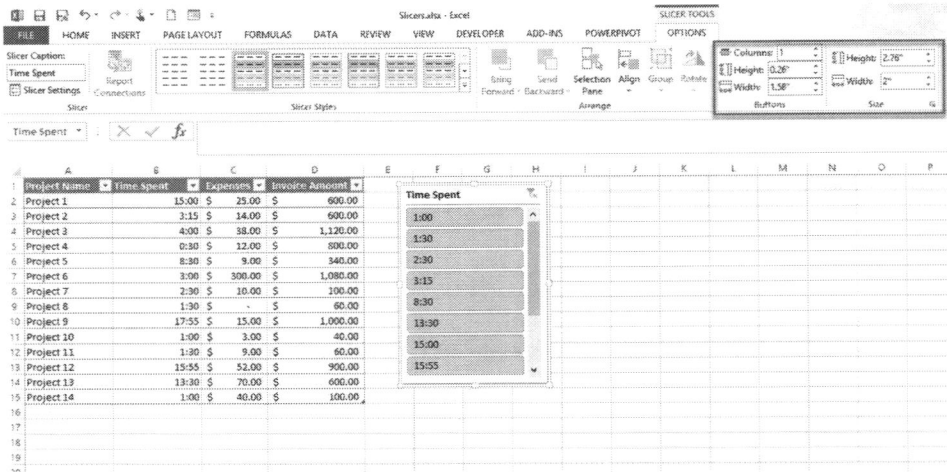

Using a Timeline

To create a timeline in an existing table, use the following procedure.

Step 1: Place your cursor anywhere in the table.

Step 2: Select the Insert tab from the Ribbon.

Step 3: Select Timeline.

Step 4: In the Insert Timelines dialog box, select the check boxes of the fields from the table for which you want to create a timeline.

Step 5: Select OK.

Step 6: Click a period on the timeline to include the date range you want to include in the PivotTable. You can also drag the edges of the timeline to expand the date range shown in the PivotTable. Remember that the timeline is interactive, so feel free to experiment.

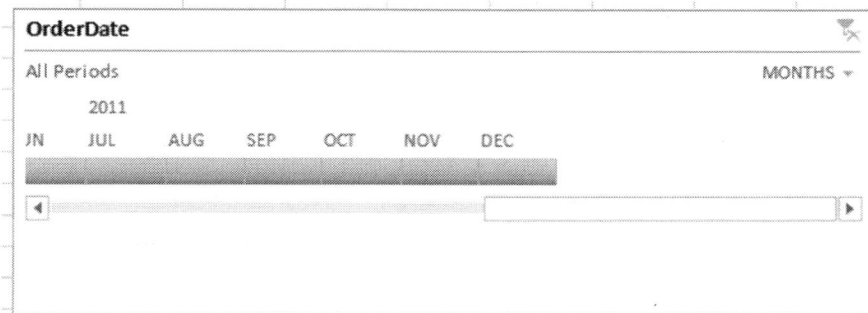

Step 7: You can change the period available in the slicer by clicking on the arrow next to the period shown (in this example, Months). Select a new period from the drop-down list.

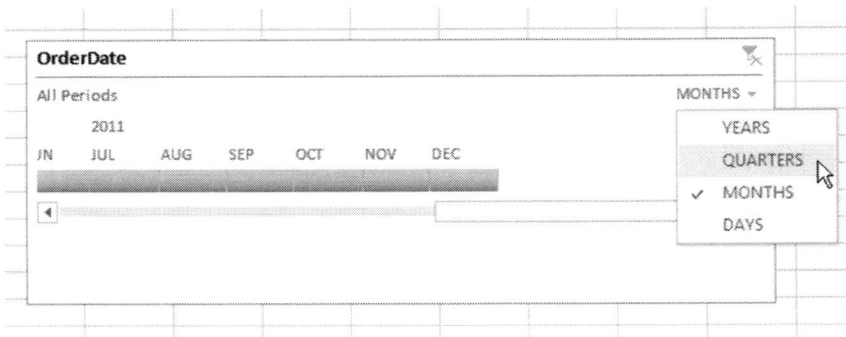

Review the Timeline Tools Options tab on the Ribbon and the Format Timeline pane. You can open the Format Timeline pane using the small square next to the Size group on the Timeline Tools Options tab on the Ribbon.

The Data Model in Excel 2016 is a new approach for integrating data from multiple tables. It builds a relational data source inside an Excel workbook, but it is virtually transparent. Data models provide the tabular data that is used in PivotTables, PivotCharts, and Power View reports. In this chapter, you will learn how to connect to a new external data source and create a PivotTable using an external data connection. You will use PivotTables to analyze data in multiple tables. You will also learn how to create relationships between tables.

Connecting to a New External Data Source

To connect to an external data source from an Access Database, use the following procedure.

Step 1: Select the Data tab from the Ribbon.

Step 2: Select the arrow under Get Data, choose From Database, and then select From Microsoft Access Database.

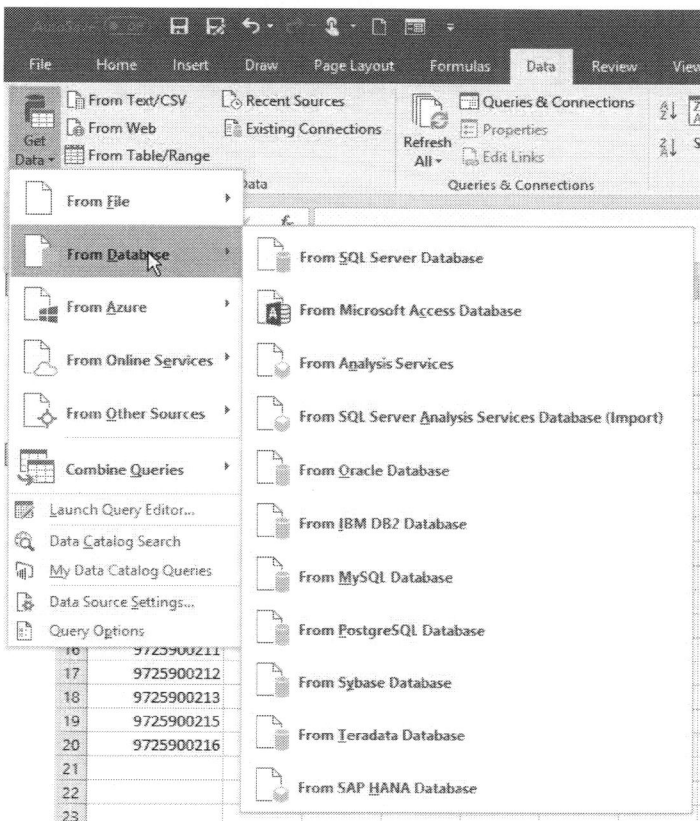

Step 3: In the Select Data Source dialog box, navigate to the location of the database you want to use and select Open.

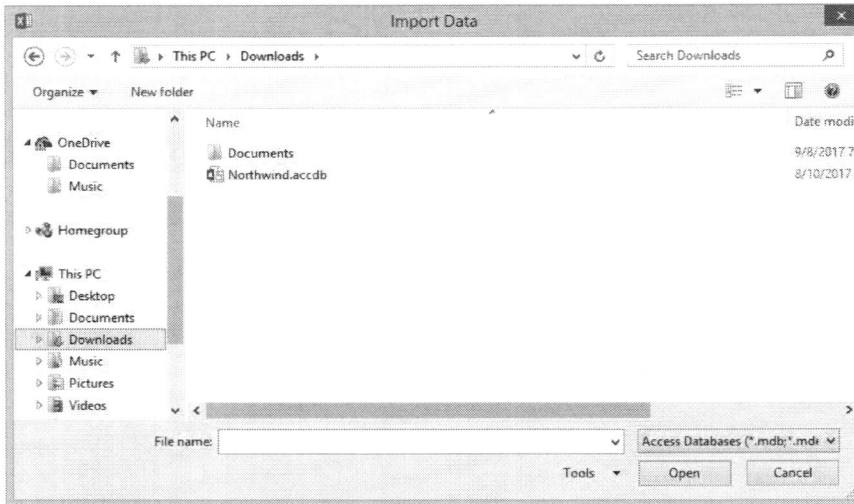

Step 4: In the Select Table dialog box, check the box for the table to which you want to connect. If you would like to select more than one table or query, check the Enable Selection of multiple tables box. You will use multiple tables later in this chapter, so include more than one table.

Step 5: Select OK.

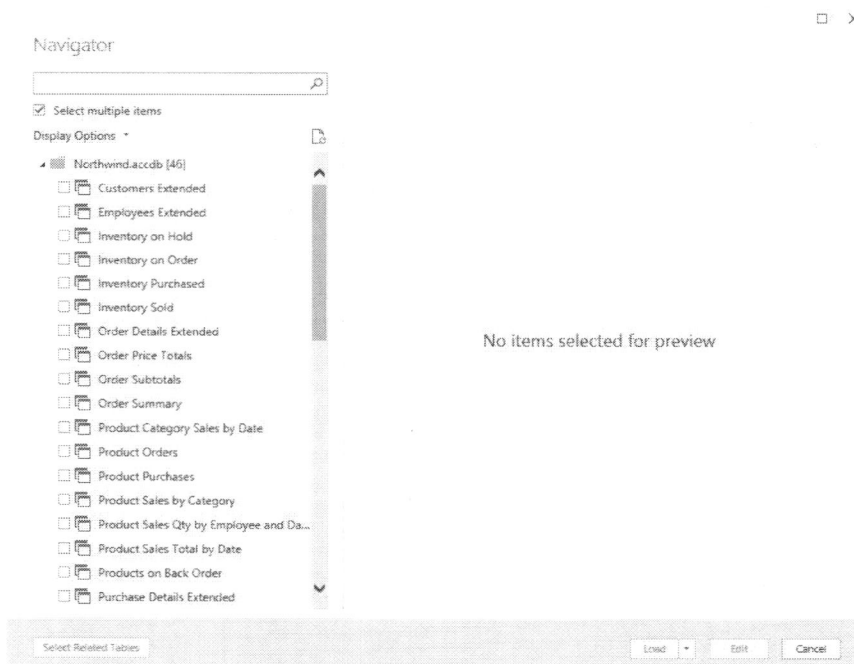

Step 6: In the Queries & Connections double-click a table to open the query editor.

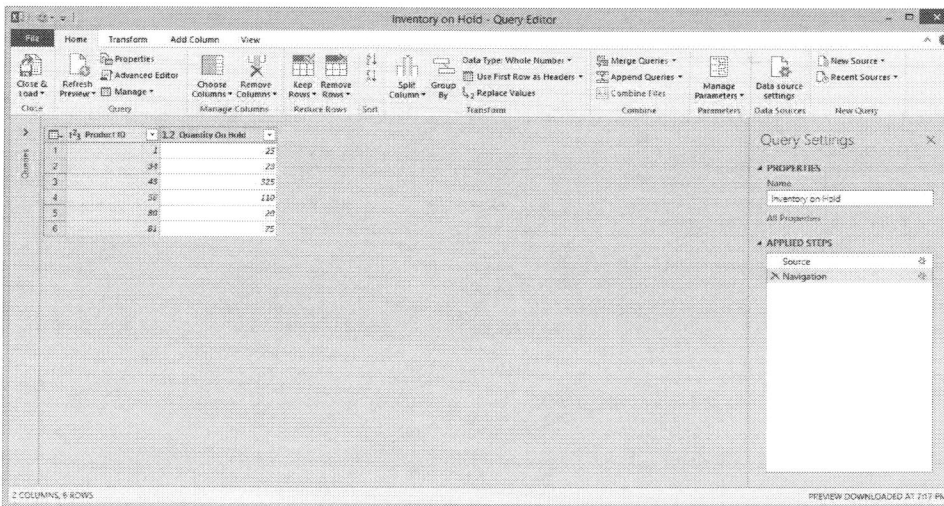

Step 7: Mouse-over the table in Queries & Connections and click View in Worksheet.

Step 8: Choose … and select Load To… to open the Import Data window.

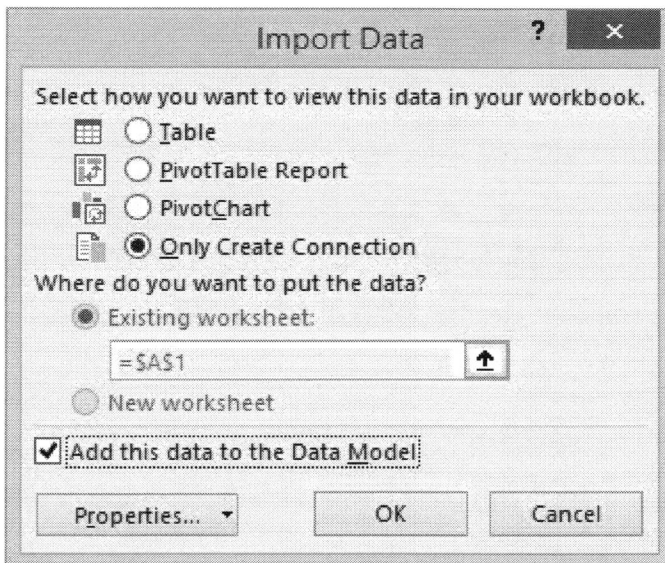

You can choose to create a connection only or a Table, PivotTable Report, PivotChart, or Power View Report. In this example, we will just create the connection.

Creating a PivotTable Using an External Data Connection

To create a PivotTable using an existing external data connection, use the following procedure.

Step 1: Select the Insert tab from the Ribbon.

Step 2: Select PivotTable.

Step 3: In the Create PivotTable dialog box, select Use an external data source.

Step 4: Select Choose Connection.

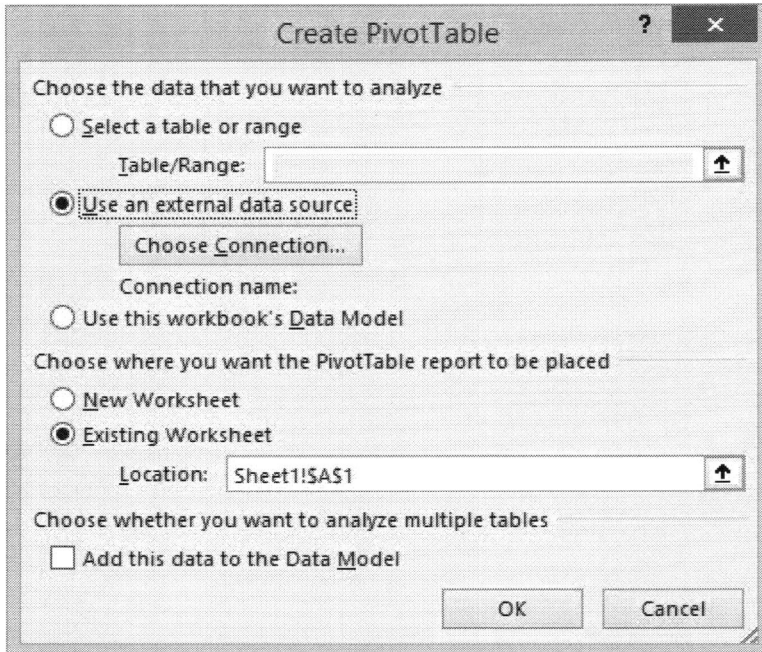

Step 5: In the Existing Connections dialog box, select the Connection that you want to use. You can use the Show drop down list to narrow the list of connections available.

Step 6: Select Open.

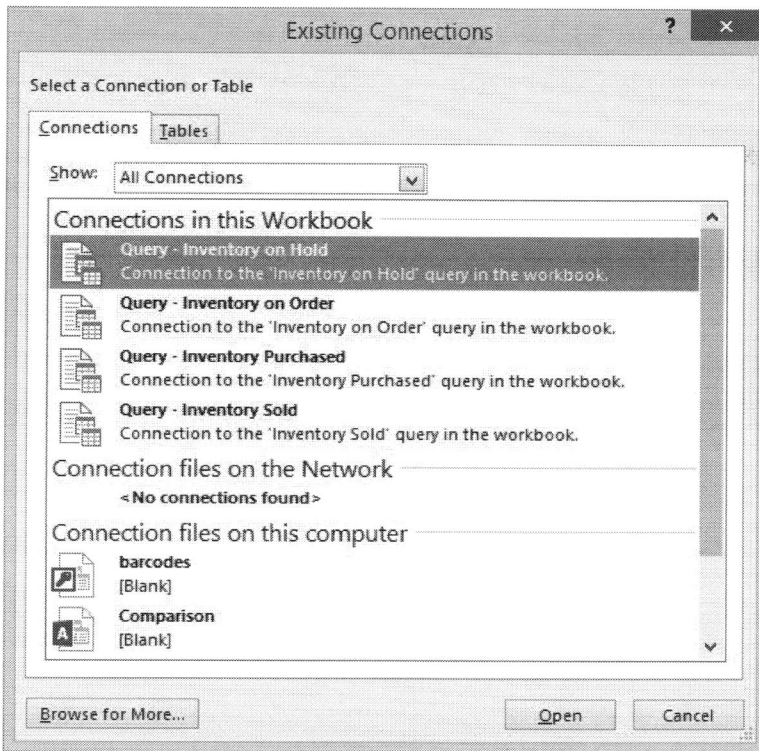

Step 7: In the Create PivotTable dialog box, choose where to place the new PivotTable.

Step 8: Select OK.

Excel adds an empty PivotTable and shows the Field pane so that you create the PivotTable you need.

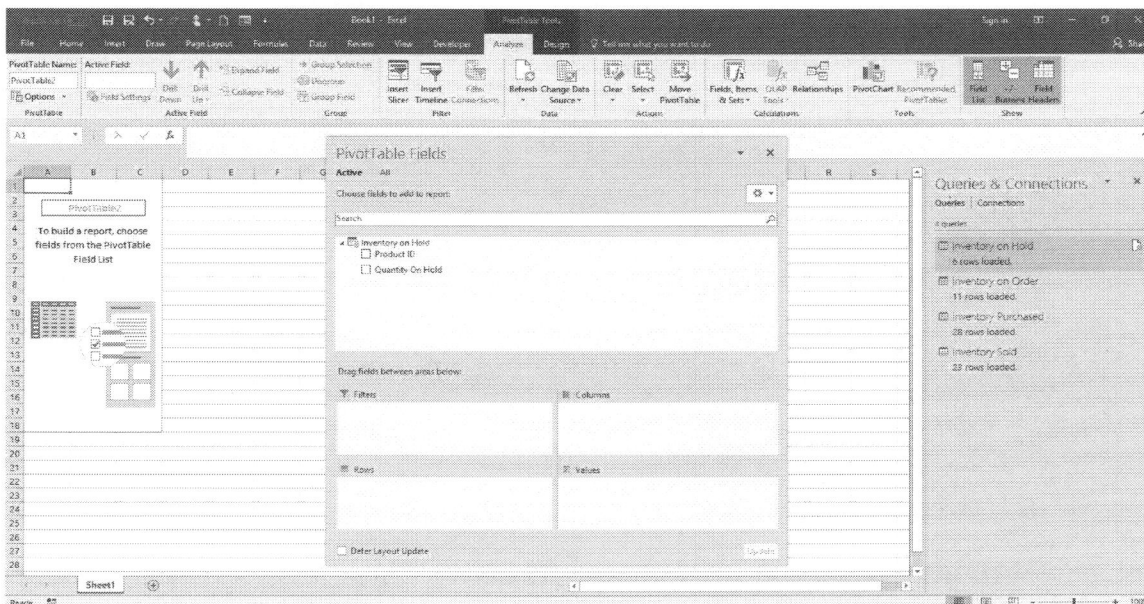

A relationship matches data in two tables to create an association. Relationships allow you to use fields from multiple tables, even when the tables originate from different sources. Each table should have a meaningful name to help you identify them when creating the relationship.

One column in one of the tables should have unique, non-duplicate data values.

To work with the PivotTable Fields pane when the PivotTable is based on multiple tables (or a data connection with multiple tables).

Step 1: The Field pane contains all the tables you selected when you imported the data. You can expand or compress the fields for each table by clicking on the plus or minus signs.

Step 2: Add the fields to the PivotTable report as you would for any PivotTable. You can add fields from any table to the VALUES, ROWS, or COLUMNS area.

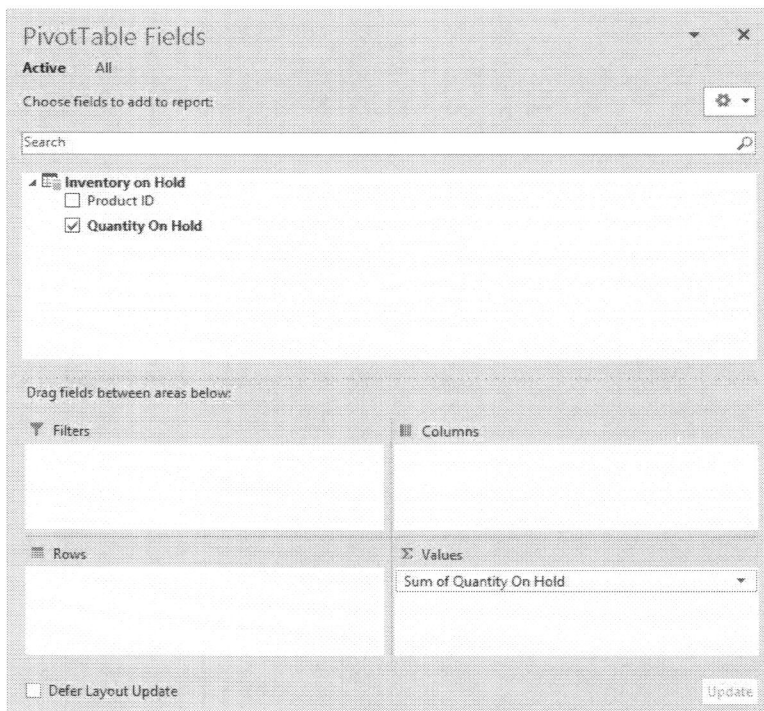

Step 3: If you select fields from tables that are not already related, Excel displays the warning: Relationships between tables may be needed. Select Create.

PivotTable Fields ▾ ✕

ACTIVE | **ALL**

Choose fields to add to report: ⚙ ▾

Relationships between tables may be needed. | CREATE... | ✕

▴ ▤ **Inventory on Hold1**
     ☑ **Product ID** ▴

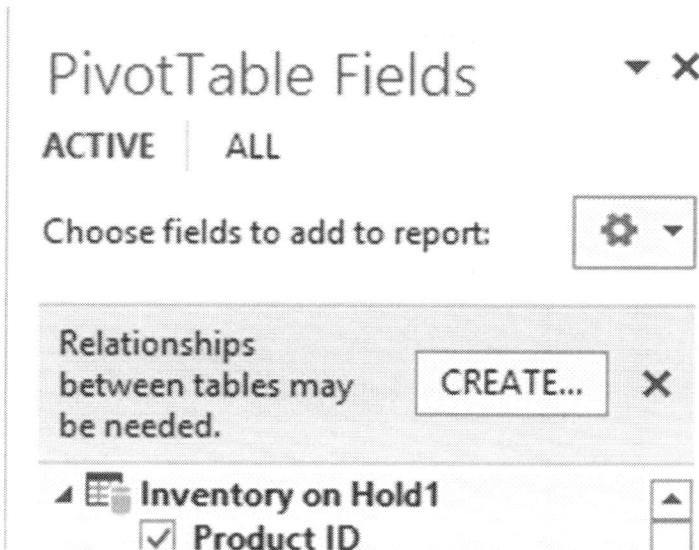

Step 4: In the Create Relationship dialog box, select the first Table from the drop-down list. In a one-to-many relationship, this table should be on the many side.

Step 5: Select the Column from that table to use in the relationship. This column should have unique values that match the values in the Related Column.

Step 6: Select the Related Table from the drop-down list. This table should have at least one column of data that is related to the table you just selected.

Step 7: Select the Column from that table to use in the relationship from the drop-down list.

Step 8: Select OK.

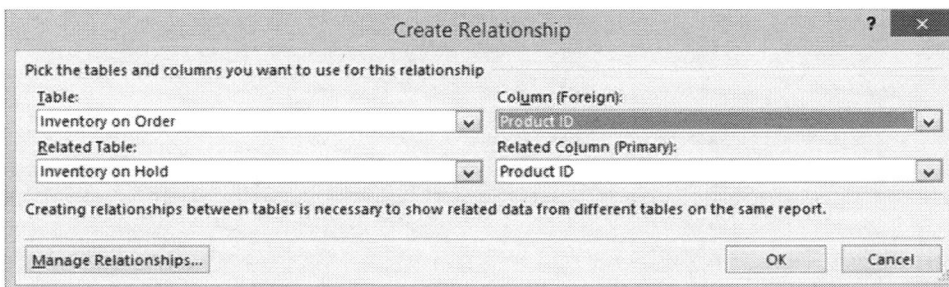

Create Relationship ? ✕

Pick the tables and columns you want to use for this relationship

Table: | Column (Foreign):
Inventory on Order ▾ | Product ID ▾

Related Table: | Related Column (Primary):
Inventory on Hold ▾ | Product ID ▾

Creating relationships between tables is necessary to show related data from different tables on the same report.

Manage Relationships... | OK | Cancel

Power View is an interactive way to explore, visualize, and present data that encourages ad-hoc reporting. This chapter will introduce you to Power View with lessons that include downloaded data from the Windows Azure Marketplace to learn how to create a Power View sheet, add additional tables to the data model used by the Power View Sheet and add an additional chart and change it to a map.

About Power View

Make sure that the PowerPivot Add-in is enabled, use the following procedure.

Step 1: Select the File tab to open the Backstage View.

Step 2: Select Options.

Step 3: Select Add-ins.

Step 4: From the Manage drop down list, select COM Add-ins.

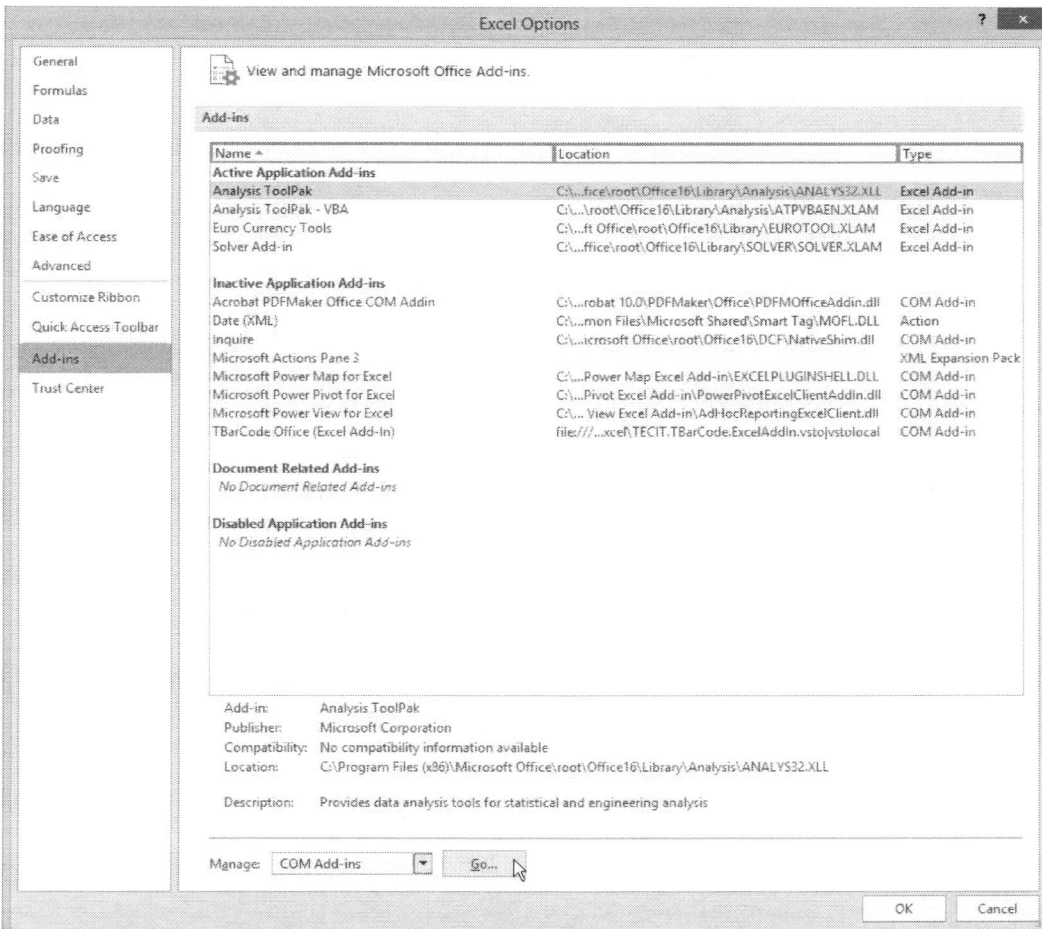

Step 5: Check the Enable PowerPivot box.

Step 6: Select OK.

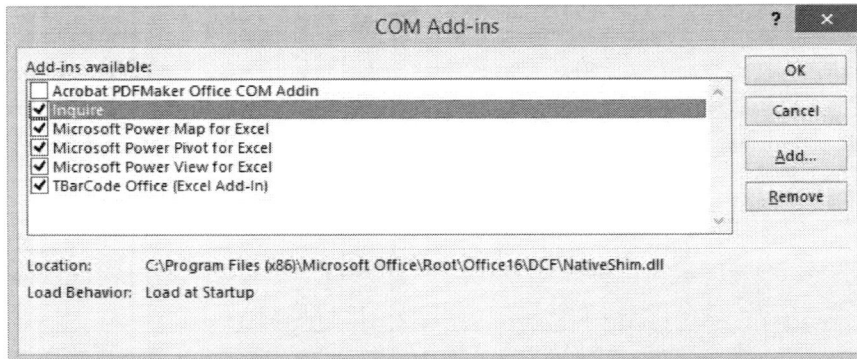

Creating a Power View Sheet

To download the sample data, which creates a data model, use the following procedure.

Step 1: Select the Power Pivot tab from the Ribbon. You may need to restart Excel after enabling the Add-in.

Step 2: Select Manage Data Model.

Step 3: In the PowerPivot for Excel window, select Get External Data. Select From Other Sources. Select From Microsoft Azure Marketplace and press Next.

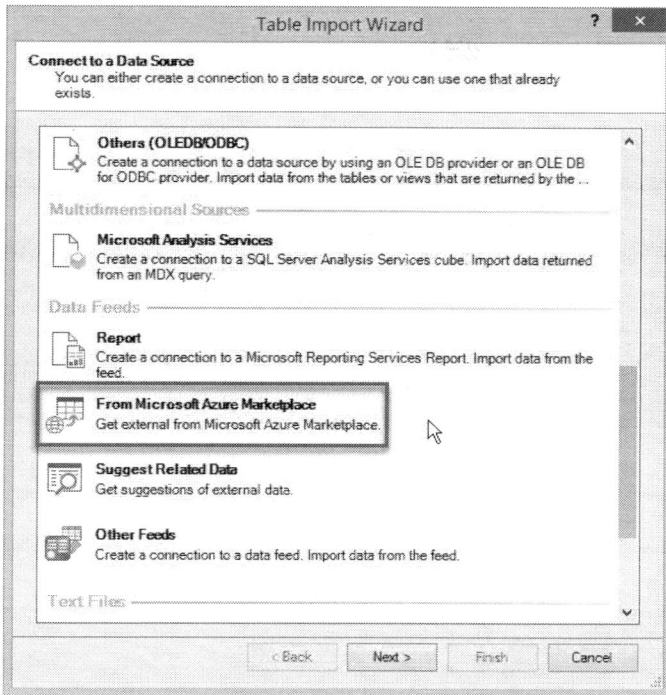

Step 4: In the Table Import Wizard window, you can filter the list of options. To find the one for this lesson, select the following filters: Free and Transportation and Navigation.

Step 5: Next to US Air Carrier Flight Delays, select Subscribe.

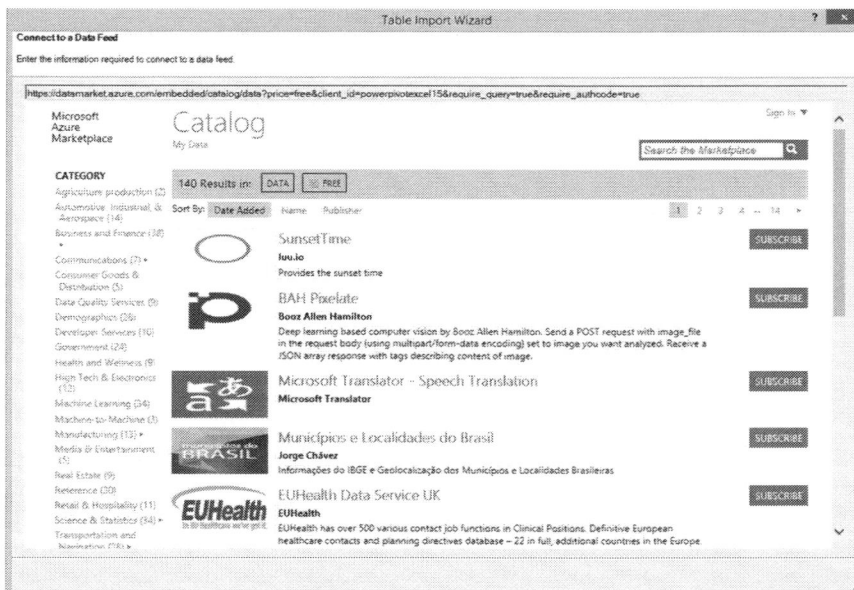

Step 6: Sign in with your Microsoft account. If necessary, create a Windows Azure Marketplace account. Continue making selections until you have finished subscribing to the information.

Step 7: At the end of the sample data, choose Select Query.

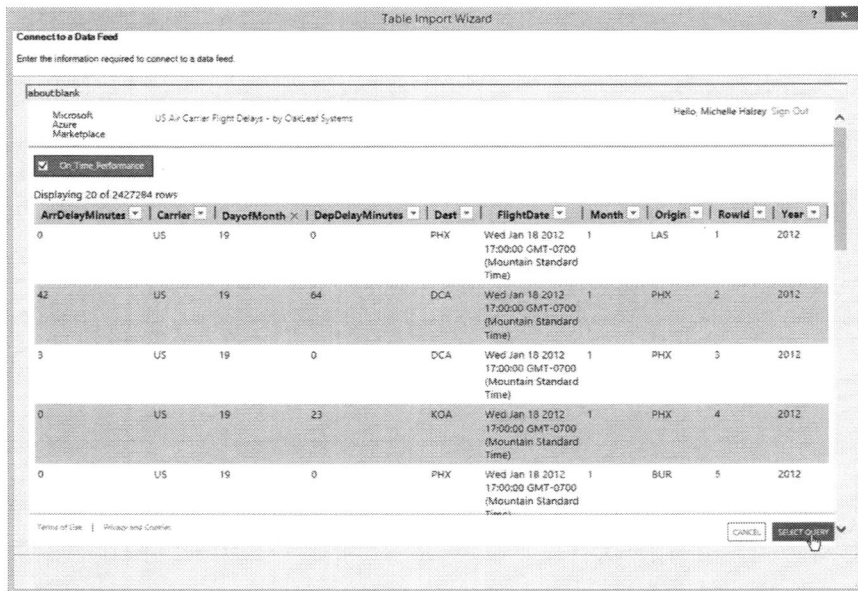

Step 8: You can change the name for the connection, if desired. Select Next.

Step 9: Select Preview & Filter from the Table Import Wizard.

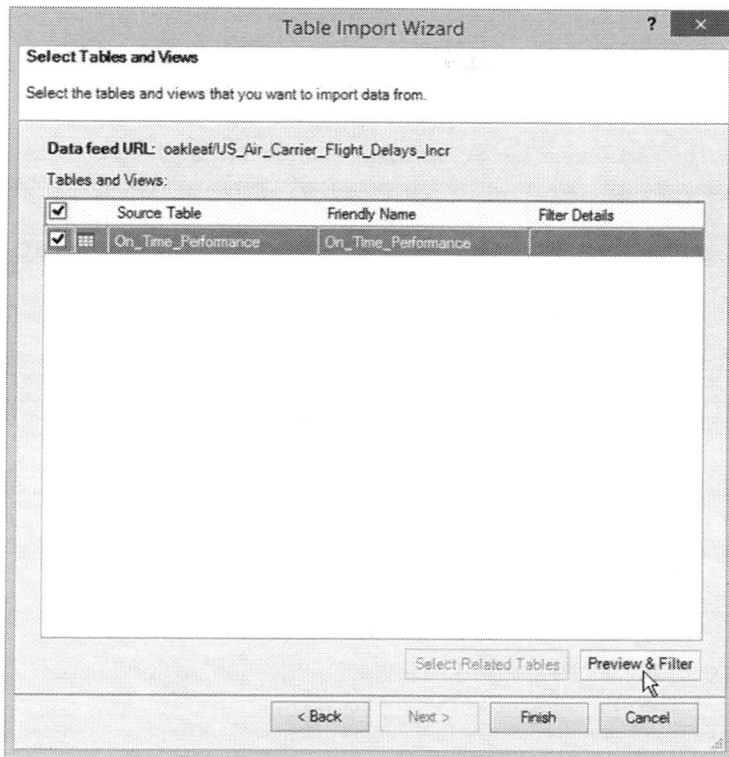

Step 10: Select OK to import all the data. Select Finish. Be aware that this may take some time.

Microsoft Excel 2016 does not have Power View turned on by default. You may need to add it back into the ribbon to use Power View.

To enable Power View

Step 1: Click he File tab of the ribbon and click Options.

Step 2: Choose Customize Ribbon.

Step 3: Click the Insert tab where you want to add the new group on the ribbon and click New Group.

Step 4: Click Rename to give the group the name "Report".

Step 5: Click Commands Not in the Ribbon under the Choose Commands From menu and then pick Step 6: Add the Insert a Power View Report command into the custom group you created.

The first time you use the Insert a Power View Report into a spreadsheet, you may see a message that the program is opening Power View. Click Enable to use the add-in.

To create a Power View sheet, use the following procedure.

Step 1: Select the Ribbon to which you added the new group.

Step 2: Select Power View Reports.

Power
View
Report

Now let us add some fields to the report.

Step 1: Expand the table.

Step 2: Check the fields to add to the reports.

Step 3: Select the desired fields in the Power View Fields window.

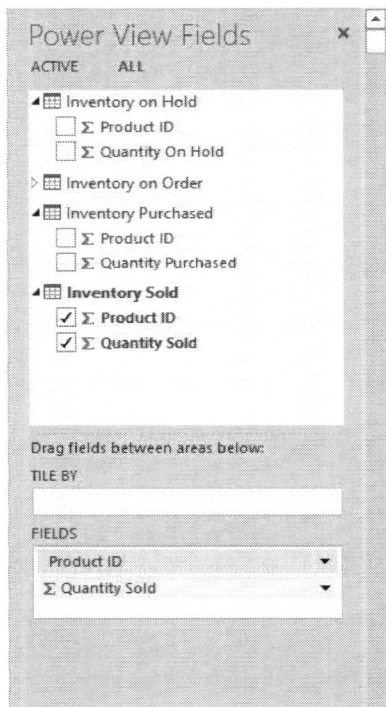

Step 4: In the Fields drop-down, select the appropriate function.

Now we will select a chart type.

Step 1: Click the table in the report to open the Design ribbon in the Table Tools options.

Step 2: Select the Design tab from the Ribbon.

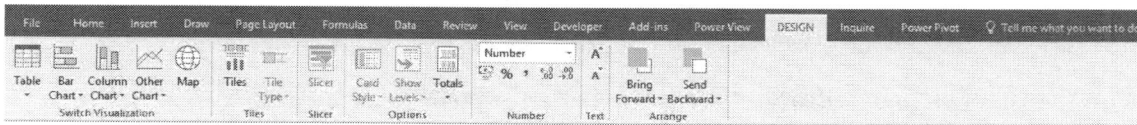

Step 3: Select Column Chart. Select Clustered.

Step 4: Let us add another field. In the Field list, drag a field to the appropriate location.

Add a Table to the Data Model

To add a table to the data model, use the following procedure.

Step 1: Go to this site: http://www.airfarewatchdog.com/pages/3799702/airline-letter-codes/.

Step 2: Copy the two columns including the codes and the airline carrier names.

Step 3: On a blank worksheet in your workbook, enter the following headings in cell A1 and B1, respectively: AirlineCode and AirlineName

Step 4: In cell A2, paste the data you copied from the website.

Step 5: Format the data as a table.

Step 6: On the Table Tools Design tab, rename the table Airlines.

Step 7: Also rename the worksheet Airlines.

Now we will relate the tables in Power View, use the following procedure.

Step 1: Return to the Power View sheet.

Step 2: Your Airlines table should be in the Field list. If not, select All from the Field list.

Step 3: Select the column chart.

Step 4: Remove Carrier from the Axis box.

Step 5: Expand the Airlines tab to check the AirlineName box.

Excel displays the "Relationships between tables may be needed" message.

Step 6: Select Create.

Step 7: Select Airlines as the first table.

Step 8: Select AirlineCode as the column.

Step 9: Select On_Time_Performance as the second table.

Step 10: Select Carrier as the column.

Step 11: The message warning about the relationship just means that a primary key is created instead of a foreign key.

Step 12: Select OK.

To get the airport code data added to the data model, use the following procedure.

Step 1: Go to this site: http://www.airportcodes.us/us-airports.htm.

Step 2: Copy the four columns, including the code, name, city and state without the table heading.

Step 3: Add a new sheet to your workbook.

Step 4: Paste the data in cell A1.

Step 5: Rename the columns so that we can relate the data.

- Code = AirportCode
- Name = AirportName
- City = AirportCity
- State = AirportState

Step 6: Format the data as a table.

Step 7: On the Table Tools Design tab, rename the table Airports.

Step 8: Also rename the worksheet Airports.

Step 9: Return to the Power View sheet.

Step 10: Remove Origin from the Axis box.

Step 11: Expand the Airports table and check the AirportName box.

Step 12: Select Create in the message to create the appropriate relationships.

Step 13: Select Airports as the first table.

Step 14: Select AirportCode as the column.

Step 15: Select On_Time_Performance as the related table.

Step 16: Select Origin as the column.

Step 17: Select OK.

To create another chart, use the following procedure.

Step 1: You can start another visualization not related to your current fields by clicking on the blank area outside the previous chart.

Step 2: In the Field list, on the On_Time_Performance table, check the Origin box. Also check the DepDelayMinutes box.

Step 3: Select the arrow next to DepDelayMinutes and select Average.

Step 4: Let us decrease the decimal for that column. Click on the column and use the Decrease Decimal tool on the Design tab of the Ribbon.

Step 5: Now let us convert this table to a bar chart. Select Bar Chart from the Design tab on the Ribbon. Select Stacked Bar.

Step 6: You can move and resize the chart to maximize what you can see with the available real estate on the sheet.

To create a map, use the following procedure.

Step 1: Select the Airports chart.

Step 2: Select the Design tab from the Ribbon.

Step 3: Select Map.

Step 4: If Power View puts AirportName in the Color box, drag it to the Locations box.

Step 5: Select the plus sign in the upper right corner of the map to zoom in. Use the cursor to reposition the map.

To filter the map, use the following procedure.

Step 1: Select the map.

Step 2: In the Filters area, select Map.

Step 3: Select Average of DepDelayMinutes.

Step 4: Drag the left side of the scroll bar to show only delays that are greater than 10 minutes.

Tell Me

Instead of searching the online help or in Excel, you can use the Tell Me feature to look for the solution you need.

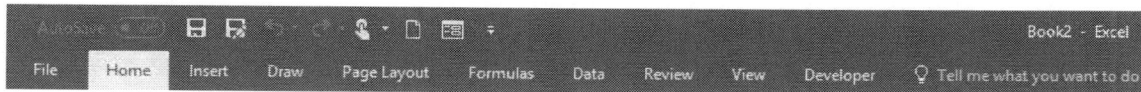

Add-ins

It is now easier to integrate Microsoft Excel Add-ins with your spreadsheet. You can download partner applications in the same manner as templates from the Office Store.

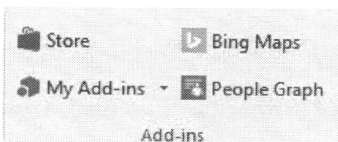

Step 1: Select the Insert ribbon.

Step 2: Click Store in the Add-ins menu.

Step 3: Search for the desired add-in in the Store and download and install the add-in.

The MY Add-ins drop-down menu will show the currently installed Add-ins from the online Microsoft Store.

Draw Ribbon

New versions of office give you the option to draw freehand notations or shapes, and gives you the ability to highlight text on the new Draw ribbon. This ribbon includes the following tools:

- Eraser – to erase any items you add to the page.
- Lasso select to select a segment of the screen
- Pens used to write or highlight
- Ink to shape to draw a shape and then have the shape snap to the diagram and fill with the pen color

- Ink to Math to insert math equations into the spreadsheet. It can also open the equation editor.

The new tools give you the option to draw with a digital pen, your finger, or a mouse. If your device is touch enabled, this ribbon is on by default. If the device is not touch enabled, you will need to turn on the ribbon in the Options menu.

Step 1: Go to the File ribbon and click Options.

Step 2: Select Customize Ribbon.

Step 3: Add a check mark next to the check box labeled Draw in the box on the right side of the dialog box.

Write, draw, or highlight text

The pen is portable and customizable. You define the pens they are then available in Microsoft Word, Microsoft Excel, and Microsoft PowerPoint.

Step 1: Tap a pen or highlighter on the Draw ribbon.

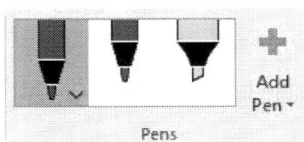

Step 2: Open the pen menu to set the Thickness and Color options for the pen. Select the preferred color and size.

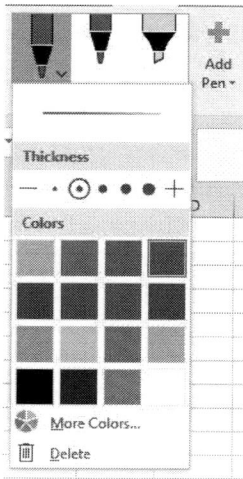

- Five predefined pen thicknesses from .25 mm to 3.5 mm. Select a thickness or use the minus or plus sign to make the pen thinner or thicker.
- Sixteen solid colors are available on the menu,
- Tap More Colors to see more options for colors

Step 3: Write or draw on the touch screen.

Once you draw an ink shape, it behaves like a shape you are used to working with in Office. You can select the shape, move or copy it, change its color, and pivot its position.

Step 4: Pick Select on the Draw tab to stop inking and select your annotations to modify or move them.

Erase ink

Step 1: Click the Draw ribbon select the Eraser in the Tools menu.

Step 2: Drag the eraser over the ink you want to remove with your pen, mouse, or finger.

Select parts of an ink drawing or written words

The Lasso Select tool in the Tools menu to specifically select objects drawn with ink. This is helpful if you have a mixture of ink and standard objects and you only want to select the ink object.

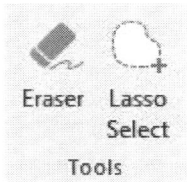

Use the Lasso Select tool to select part of a drawing or words written in ink.

Step 1: Tap Lasso Select in the Tools menu of the Draw ribbon.

Step 2: Drag a circle around the segment of the drawing or word you want to select. A faded, dashed selection region appears around it, and when you're done, the portion you lassoed is selected. Once lassoed, then you can manipulate that object.

Get & Transform

Microsoft Excel 2016 introduced Get & Transform, a powerful set of tools based on Power Query technology. These tools enable you to easily connect, combine, and shape data coming from a variety of sources. Excel's data import functionality has been enhanced and rearranged on the Data tab.

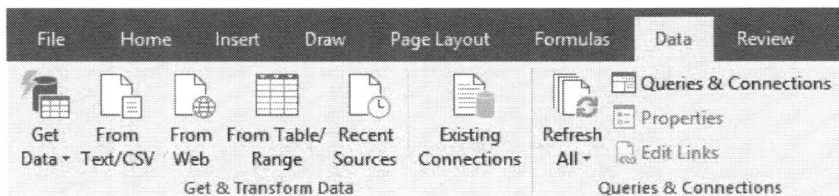

Easily connect and get data from a variety of popular data sources such as files, databases, Azure and Online services, and more. This replaces the older data import Wizards that were available under the Data ribbon.

Get Data from Text/CSV

Step 1: Click the Data ribbon.

Step 2: Click from Text/CSV in the Get & Transform Data menu.

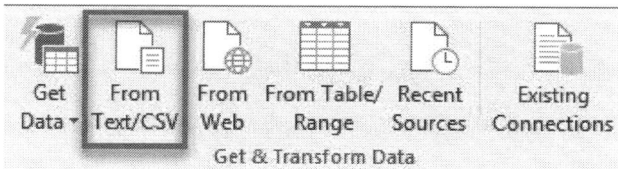

Step 3: Choose the source file and then click Ok. The connector automatically analyzes and applies import settings. If the preview is not displayed correctly, you can configure the basic import settings.

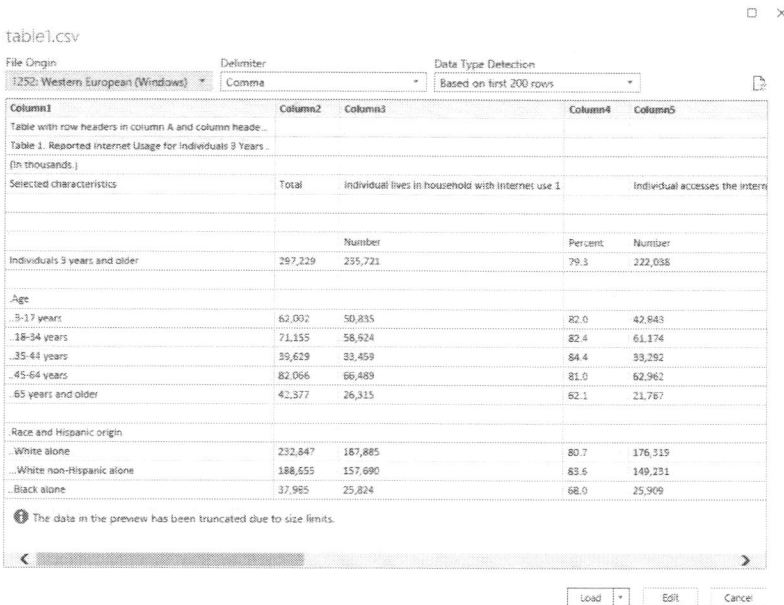

Get Data from the Web

Use the Get Data from the web connector to scrape data from HTML web pages. This replaces the legacy From Web wizard that you had on the Data ribbon.

Step 1: Click the Data ribbon.

Step 2: Click from Web in the Get & Transform Data menu.

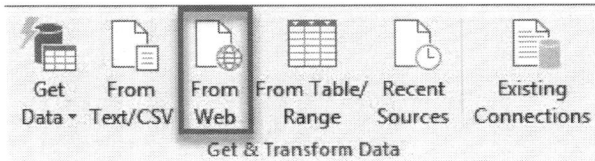

Step 3: Enter the page URL and then click Ok.

A list of the tables on the web page is displayed in the Navigator dialog. You can use Table View or Web View to interact with the web page. In Table View, click a table name on the left, and data from the table will display on the right. In Web View, you can click a table in the Navigator list, or in the Web View preview pane. The option to select multiple tables is available in either view.

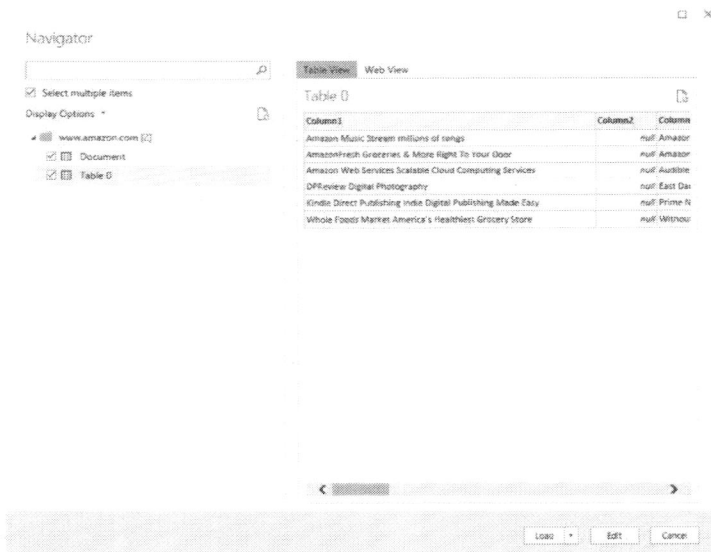

Get Data from a Table or Range

Create a new query that is linked to a table or named range in your Excel worksheet. Once added, you can then refine the data, and apply additional transformations to using the Query Editor window.

Step 1: Click the Data tab.

Step 2: Click from Table/Range in the Get & Transform Data menu.

Step 3: Modify the data using the Query Editor.

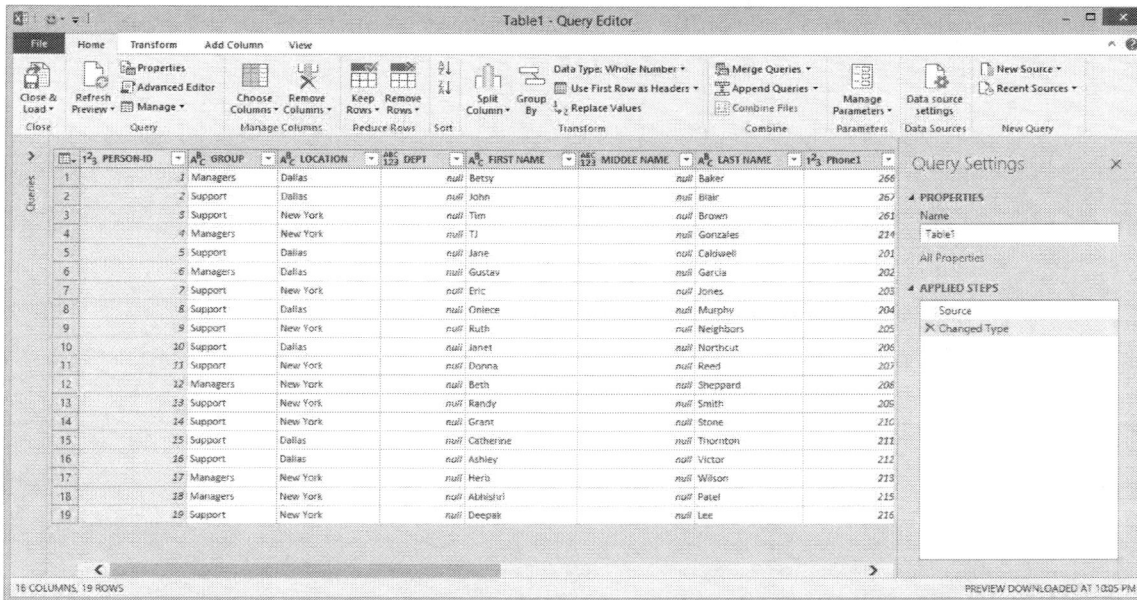

Get Data from Additional Sources

Create a new query to import data from a single data source, such as a Microsoft Access database, or you can import data from multiple text files in a folder at once.

Step 1: Click the Data ribbon.

Step 2: Select the arrow below Get Data in the Get & Transform Data menu.

Step 3: Choose the Data source.

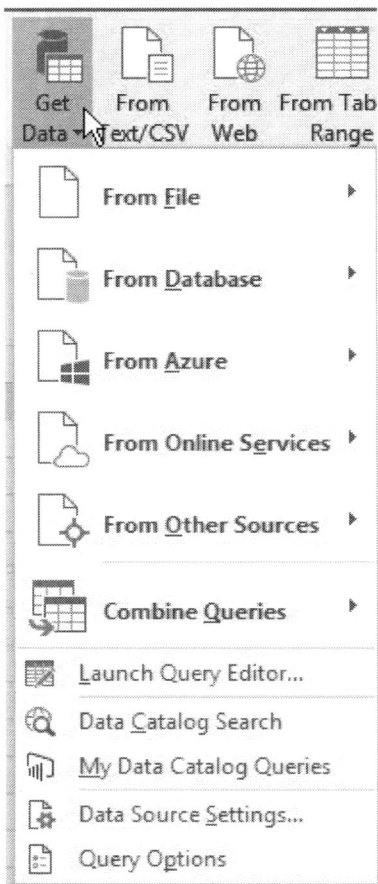

Step 4: Follow the wizard to import the data source.

The source options are organized in categories:

- From File

- From Database

- From Azure

- From Online Services

- From Other Sources

- Combine Queries

The list of available connectors is continuously growing.

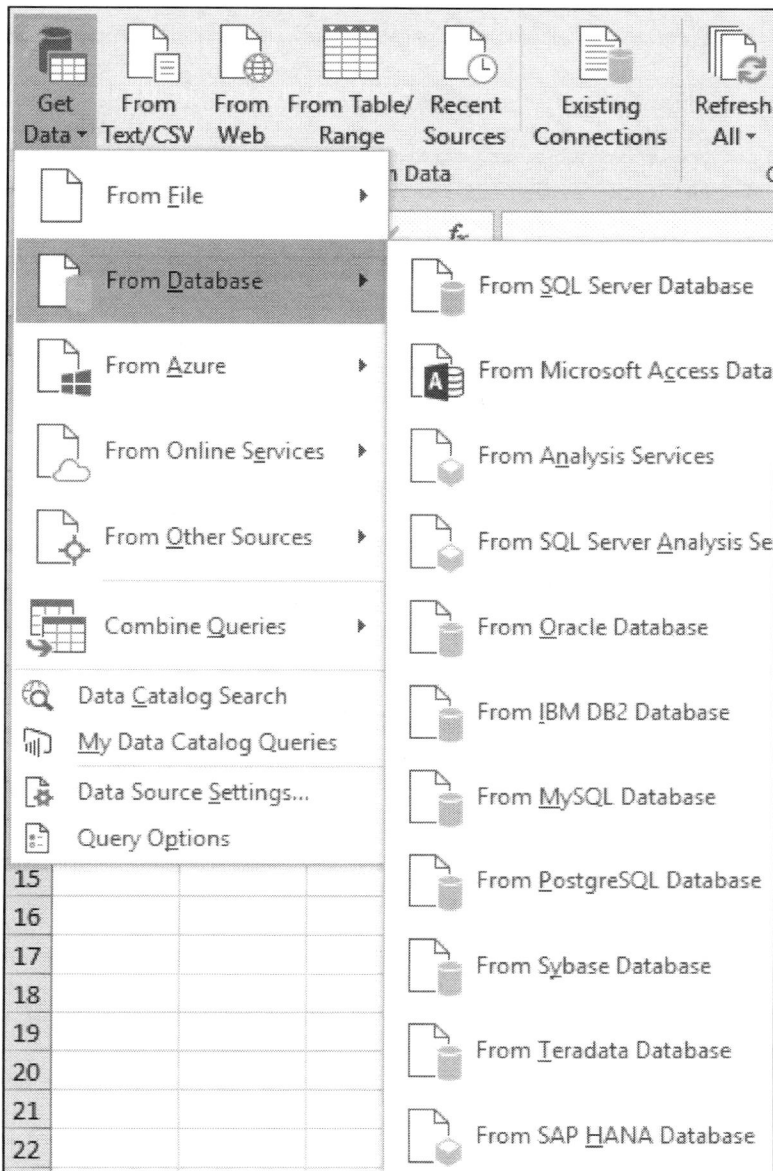

Clean & Shape Data

Use Get & Transform to clean and shape data to perform robust data analysis. Use the Query Editor window to transform the data.

Step 1: Click the Data ribbon.

Step 2: Click Get Data and connect to the desired data source and preview data in the Navigator window.

Step 3: Select the tables you would like to import and click Edit to launch the Query Editor window to apply transformations like sorting and filtering data, converting data types, splitting or merging columns, aggregating your data and more.

Step 4: Use the Applied Steps section of the Query Editor window to see the recorded transformations.

Get Data from Recent Sources

Step 1: Click the Data ribbon.

Step 2: Click Recent Sources in the Get & Transform Data menu.

Step 3: Browse to recently connected sources, select one, and then click Connect.

Restore Legacy Get External Data Option

Step 1: Click the File ribbon.

Step 2: Click Options and select the Data menu.

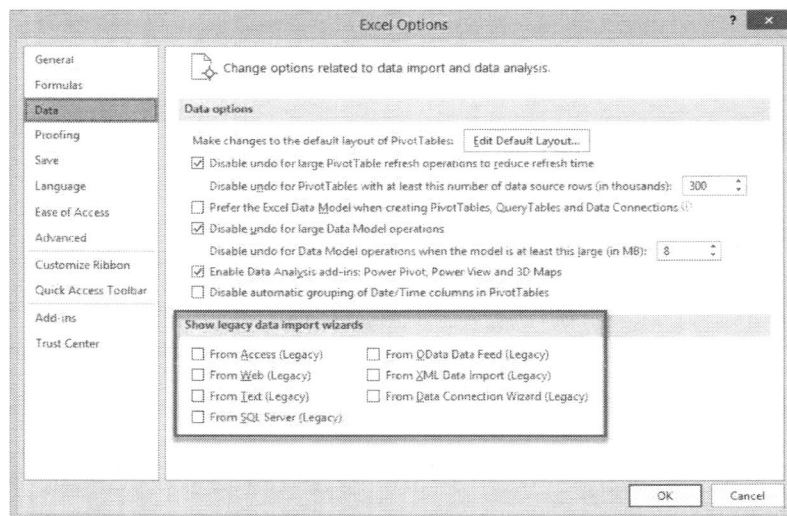

Step 3: Select the desired legacy data import options.

Step 4: A Legacy Wizards category will be added to the Get Data options.

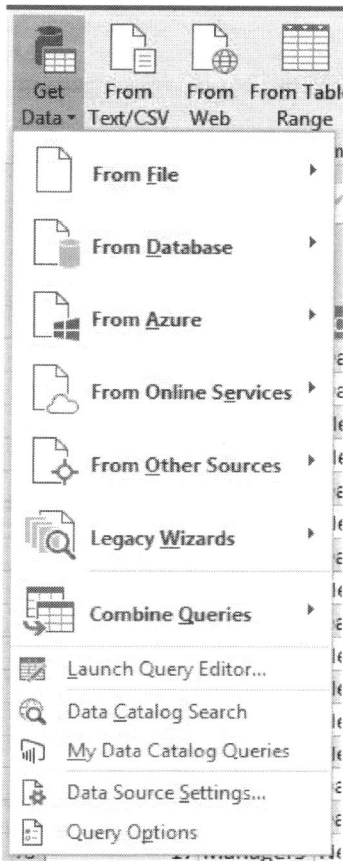

Keep Copy

In Microsoft Excel 2016, you can now copy a cell and do other functions like typing or inserting cells before pasting.

Icons

Insert an icon that displays an idea of concept using the Icons function in the Illustration menu on the Insert ribbon.

Choose and category and then an icon that represents the concept you are trying to convey.

Insert Icon

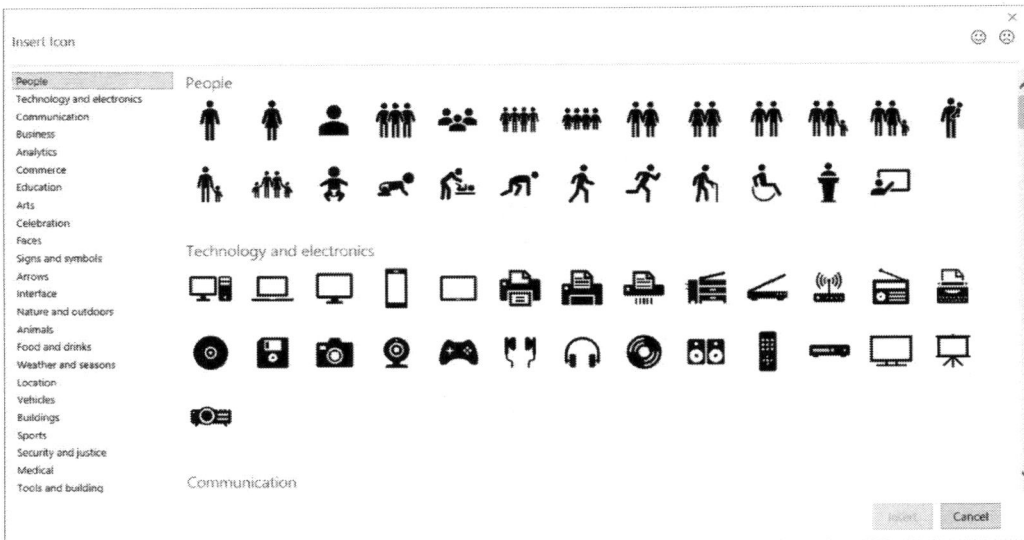

People
- People
- Technology and electronics
- Communication
- Business
- Analytics
- Commerce
- Education
- Arts
- Celebration
- Faces
- Signs and symbols
- Arrows
- Interface
- Nature and outdoors
- Animals
- Food and drinks
- Weather and seasons
- Location
- Vehicles
- Buildings
- Sports
- Security and justice
- Medical
- Tools and building

Technology and electronics

Communication

Insert   Cancel

Chapter 32 – Microsoft Excel 2016 Shortcuts

Data, Formulas, and Functions

To do this	Press
Define a name to use in references	ALT + M, M, D
Display menu or message for Error Checking button	ALT + SHIFT + F10
displays the **Control** menu for the Microsoft Office Excel window.	ALT + SPACEBAR
Copies a formula from the cell above the active cell into the cell or the Formula Bar.	CTRL + '
Switch between displaying cell values or formulas in the worksheet.	CTRL + ` (grave accent)
Copy formula from cell above the active cell into the cell or the Formula Bar	CTRL + ' (apostrophe)
Display Function Arguments dialog box when insertion point is to right of a function name in a formula	CTRL + A
Calculate all worksheets in all open workbooks, regardless of if they have changed	CTRL + ALT + F9
Check dependent formulas, and then calculate all cells in all open workbooks, including cells not marked as needing to be calculated.	CTRL + ALT + SHIFT + F9
Invoke Flash Fill to automatically recognize patterns in adjacent columns and fill current column	CTRL + E
Move cursor to end of text when in formula bar	CTRL + END
Copy value from cell above the active cell into cell or formula bar	CTRL + SHIFT + " (straight quotation mark)

To do this	Press
Insert argument names and parentheses when insertion point is to right of a function name in a formula	CTRL + SHIFT + A
Select all text in formula bar from cursor position to end	CTRL + SHIFT + END
Expand or collapse formula bar	CTRL + SHIFT + U
Create a chart of data in current range in a separate Chart sheet	F11
If editing is turned off for the cell, move the insertion point into the formula bar.	F2
Calculate all worksheets in all open workbooks	F9
Calculate active worksheet	SHIFT + F9
Inserts the argument names and parentheses when the insertion point is to the right of a function name in a formula.	CTRL + SHIFT + A
Create embedded chart of data in current range	ALT + F1

Editing

To do this	Press
Deletes one character to the left in the Formula Bar. Also clears the content of the active cell. In cell editing mode, it deletes the character to the left of the insertion point.	BACKSPACE
Displays the **Delete** dialog box to delete the selected cells.	CTRL + - (minus)

To do this	Press
Enters current date.	CTRL + ; (semi-colon)
Uses the **Undo** or **Redo** command to reverse or restore the last automatic correction when AutoCorrect Smart Tags are displayed.	CTRL + SHIFT + Z
Repeat last command or action, if possible.	CTRL + Y
Undo last action	CTRL + Z
Removes the cell contents (data and formulas) from selected cells without affecting cell formats or comments. In cell editing mode, it deletes the character to the right of the insertion point.	DELETE
Completes a cell entry from the cell or the Formula Bar, and selects the cell below (by default). In a data form, it moves to the first field in the next record. Opens a selected menu (press F10 to activate the menu bar) or performs the action for a selected command. In a dialog box, it performs the action for the default command button in the dialog box (the button with the bold outline, often the OK button).	ENTER
Displays the **Paste Name** dialog box. Available only if names have been defined in the workbook.	F3
Repeats the last command or action, if possible. Cycle through all combinations of absolute and relative references in a formula if a cell reference or range is selected.	F4

To do this	Press
Opens the **Format Cells** dialog box with the **Font** tab selected.	CTRL + SHIFT + F5
Center align cell contents	ALT + H, A, C
Add borders	ALT + H, B
Choose a fill color	ALT + H, H
Delete column	ALT + H,D, C
Open the Formulas tab	ALT + M
Hides the selected columns.	CTRL + 0
Open Format Cells dialog box	CTRL + 1
Applies or removes bold formatting.	CTRL + 2
Applies or removes italic formatting.	CTRL + 3
Applies or removes underlining.	CTRL + 4
Apply or remove strikethrough formatting	CTRL + 5
Display or hide outline symbols	CTRL + 8
Displays the **Paste Special** dialog box. Available only after you have cut or copied an object, text, or cell contents on a worksheet or in another program.	CTRL + ALT + V
Bold text or remove bold formatting	CTRL + B or CTRL + 2
Copy selected cells	CTRL + C
Use the Fill Down command to copy the contents	CTRL + D

To do this	Press
and format of the topmost cell of a selected range into the cells below.	
Applies or removes italic formatting.	CTRL + I
Italicize text or remove italic formatting	CTRL + I or CTRL + 3
Open Insert hyperlink dialog	CTRL + K
Display Create Table dialog box	CTRL + L or CTRL + T
Display Quick Analysis options for selected cells that contain data	CTRL + Q
Uses the **Fill Right** command to copy the contents and format of the leftmost cell of a selected range into the cells to the right.	CTRL + R
Apply Number format with two decimal places, thousands separator, and minus sign (-) for negative values	CTRL + SHIFT + ! (exclamation point)
Apply Date format with the day, month, and year	CTRL + SHIFT + # (number or pound sign)
Apply Currency format with two decimal places (negative numbers in parentheses)	CTRL + SHIFT + $ (dollar sign)
Apply Percentage format with no decimal places	CTRL + SHIFT + % (percent)
Apply outline border to selected cells	CTRL + SHIFT + & (ampersand)
Unhides any hidden rows within the selection.	CTRL + SHIFT + ((left parenthesis)

To do this	Press
Unhides any hidden columns within the selection.	CTRL + SHIFT +) (right parenthesis)
Enter current time	CTRL + SHIFT + : (colon)
Apply the Time format with hour and minute, and AM or PM	CTRL + SHIFT + @ (at sign)
Apply Scientific number format with two decimal places	CTRL + SHIFT + ^ (caret)
Remove outline border from selected cells	CTRL + SHIFT + _ (underline)
Apply General number format	CTRL + SHIFT + ~ (tilde)
Open Insert dialog to insert blank cells	CTRL + SHIFT + + (plus)
Format fonts in Format Cells dialog box	CTRL + SHIFT + F or CTRL + SHIFT + P
Underline text or remove underline	CTRL + U or CTRL + 4
Inserts the contents of the Clipboard at the insertion point and replaces any selection. Available only after you have cut or copied an object, text, or cell contents.	CTRL + V
Cut selected text	CTRL + X
Edit active cell and put the insertion point at the end of cell contents	F2
Add or edit a cell comment	SHIFT + F2

Help

To do this	Press
Open **Tell me** box to search for Help	ALT + Q, and then enter the search term.
Displays the **Excel Help** task pane.	F1
Turns key tips on or off. (Pressing Alt does the same thing.)	F10

Macros & VBA

To do this	Press
opens the Microsoft Visual Basic For Applications Editor, in which you can create a macro by using Visual Basic for Applications (VBA).	ALT + F11
Displays macro dialog box to create, run, edit, or delete a macro.	ALT + F8

Make Selections and Perform Actions

To do this	Press
Start a new line in same cell	ALT + ENTER
Selects the entire worksheet.	CTRL + A or CTRL + SHIFT + Spacebar
If the worksheet contains data, selects the current region. Pressing a second time selects the current region and its summary rows. Pressing third time selects the entire worksheet.	CTRL + A or CTRL + SHIFT + Spacebar
When the insertion point is to the right of a function name in a formula, displays the **Function Arguments** dialog box.	CTRL + A

To do this	Press
Fill selected cell range with current entry	CTRL + ENTER
Select current region around active cell or select entire PivotTable report.	CTRL + SHIFT + * (asterisk)
Extend selection of cells to last nonblank cell in same column or row as the active cell, or to next nonblank cell.	CTRL + SHIFT + ARROW KEY
Extend the selection of cells to the last used cell on the worksheet (lower-right corner).	CTRL + SHIFT + END
Extend selection of cells to beginning of worksheet	CTRL + SHIFT + Home
Selects all cells that contain comments.	CTRL + SHIFT + O
Opens the **Format Cells** dialog box with the **Font** tab selected.	CTRL + SHIFT + P
Select current and next sheet in a workbook	CTRL + SHIFT + Page Down
Selects the current and previous sheet in a workbook. In a dialog box, performs the action for the selected button, or selects or clears a check box.	CTRL + SHIFT + PAGE UP
Select all objects on worksheet when an object is selected	CTRL + SHIFT + Spacebar
Select an entire column in a worksheet	CTRL + Spacebar
Displays the **Create Table** dialog box.	CTRL + T
Selects the next or previous command when a menu or submenu is open. When a Ribbon tab is selected, these keys navigate up or down the tab group. In a dialog box, arrow keys move between options in an open drop-down list, or between options in a group	DOWN ARROW or UP ARROW

To do this	Press
of options.	
Turns extend mode on or off. In extend mode, **Extended Selection** appears in the status line, and the arrow keys extend the selection.	F8
Turn extend mode on and use arrow keys to extend a selection Press again to turn off	F8 + Arrow Keys + F8
Select first command on menu when menu or submenu is visible	Home
Extend selection of cells by 1 cell	SHIFT + arrow key
Completes a cell entry and selects the cell above.	SHIFT + ENTER
Add a non-adjacent cell or range to a selection of cells by using arrow keys	SHIFT + F8
Select an entire row in a worksheet	SHIFT + Spacebar

To do this	Press
Move one cell up, down, left, or right in a worksheet.	ARROW KEYS
Move to edge of current data region in worksheet	CTRL + arrow key
Move to the last cell on a worksheet, to the lowest used row of the rightmost used column.	CTRL + END
Move to next sheet in workbook. Switches between worksheet tabs, from right-to-left.	CTRL + Page Down
Move to previous sheet in workbook. Switches between worksheet tabs, from left-to-right.	CTRL + Page Up
Move 1 cell down in a worksheet.	Down Arrow key
Move down, up, left, or right	Down Arrow, Up Arrow, Left Arrow, or Right Arrow
Move to next nonblank cell in the same column or row as the active cell or if cells are blank, move to the last cell in the row or column.	END+ arrow key
Move to cell in upper-left corner of window when Scroll Lock is on	HOME + Scroll Lock
Move 1 cell left in a worksheet.	Left Arrow key
Move 1 screen down in worksheet	Page Down
Move 1 screen up in worksheet	Page Up
Move 1 cell right in a worksheet	Right Arrow key
Move to previous cell in a worksheet	SHIFT + Tab
Move 1 cell to the right in a worksheet or in a	Tab

To do this	Press
protected worksheet, move between unlocked cells.	
Move 1 cell up in a worksheet.	Up Arrow key

To Do This	Key
If editing a formula, toggle Point mode off or on so you can use arrow keys to create a reference.	F2
Displays the **Save As** dialog box	F12
Open the Data tab	ALT + A
Open a menu for a selected button	ALT + Down Arrow
Go to Backstage view	ALT + F
Closes Excel	ALT + F4
Open Home tab	ALT + H
Open the Insert tab	ALT + N
Open the Page Layout tab	ALT + P
Move 1 screen to right in worksheet	ALT + Page Down
Move 1 screen to left in worksheet	ALT + Page Up
Open the Review tab	ALT + R
Inserts a new worksheet.	ALT + SHIFT + F1
Open **View** tab	ALT + W
Select the active tab of the ribbon, use the access keys or the arrow keys to move to a different tab.	Alt or F10
Switch between hiding objects, displaying objects, and displaying placeholders for objects	CTRL + 6
If the worksheet contains data, selects the current region. Pressing a second time selects the current region and its summary rows. Pressing a third time	CTRL + SHIFT + SPACEBAR

To Do This	Key
selects the entire worksheet.	
switches to the previous tab in a dialog box.	CTRL + SHIFT + TAB
Closes the selected workbook window.	CTRL + W
Open a list for a selected command, when in the command	Down Arrow key
Move to next command when a menu or submenu is open	Down Arrow key
Opens a selected drop-down list.	DOWN ARROW or ALT+DOWN ARROW
Moves to the cell in the lower-right corner of the window when SCROLL LOCK is turned on. Also selects the last command on the menu when a menu or submenu is visible.	END
Selects the first command on the menu when a menu or submenu is visible.	HOME
Move to a submenu when the main menu is open or selected	Left Arrow key
Moves to the cell in the upper-left corner of the window	SCROLL LOCK ON + HOME
Open context menu (Displays shortcut menu for selected item)	SHIFT + F10 Context Key
Repeats the last Find action	SHIFT + F4
Displays the find tab	SHIFT + F5
Switches between the worksheet, Zoom controls,	SHIFT + F6

To Do This	Key
task pane, and ribbon.	
Move to previous option in a dialog box.	SHIFT + Tab
Activate selected button	Spacebar or ENTER
Move the focus to commands on the ribbon.	Tab or SHIFT + Tab
Selects the tab to the left or right when the Ribbon is selected. When a submenu is open or selected, these arrow keys switch between the main menu and the submenu. When a Ribbon tab is selected, these keys navigate the tab buttons.	LEFT ARROW or RIGHT ARROW
Displays the **Go To** dialog box.	CTRL + G
Displays the **Go To** dialog box.	F5
Moves to the beginning of a row in a worksheet.	HOME
Hide the selected rows	CTRL + 9
Displays the Clipboard.	CTRL + C + CTRL + C
Fills the selected cell range with the current entry.	CTRL + ENTER
Displays or hides the ribbon, a component of the Microsoft Office Fluent user interface.	CTRL + F1
Maximizes or restores the selected workbook window.	CTRL + F10
Displays the print preview area on the **Print** tab in the Backstage view.	CTRL + F2
Closes the selected workbook window.	CTRL + F4
Restores the window size of the selected workbook	CTRL + F5

To Do This	Key
window.	
Switches to the next workbook window when more than one workbook window is open.	CTRL + F6
Performs the **Move** command on the workbook window when it is not maximized. Use the arrow keys to move the window, and when finished press Enter, or Esc to cancel.	CTRL + F7
Performs the Size command when a workbook is not maximized.	CTRL + F8
Minimizes a workbook window to an icon.	CTRL + F9
Displays the **Find and Replace** dialog box, with the **Replace** tab selected.	CTRL + H
Move to beginning of worksheet	CTRL + Home
Creates a new, blank workbook.	CTRL + N
Open a spreadsheet, display open dialog box or find a file	CTRL + O
Displays the **Print** dialog box.	CTRL + P
Saves the active file with its current file name, location, and file format.	CTRL + S
Switches to the next tab in dialog box.	CTRL + TAB
Cancels an entry in the cell or Formula Bar.	
Closes an open menu or submenu, dialog box, or message window.	ESC
It also closes full screen mode when this mode has	

To Do This	Key
been applied, and returns to normal screen mode to display the Ribbon and status bar again.	
Switches between the worksheet, ribbon, task pane, and Zoom controls. In a worksheet that has been split, F6 includes the split panes when switching between panes and the ribbon area.	F6

Made in the USA
Middletown, DE
15 June 2018